Motor-Klassiker
Engine Classics

MOTOR-KLASSIKER
ENGINE CLASSICS

Herzstücke der großen Autolegenden
Hearts of the big automobile legends

FRANZIS

Bibliografische Information der Deutschen Bibliothek

Bibliographic information published by the Deutsche Nationalbibliothek

Die Deutsche Bibliothek verzeichnet diese Publikation in der Deutschen Nationalbibliografie; detaillierte Daten sind im Internet über http://dnb.ddb.de abrufbar.

The Deutsche Nationalbibliothek lists this publication in the Deutsche Nationalbibliografie; detailed bibliographic data are available on the internet at http://dnb.dnb.de.

© 2017 Franzis Verlag GmbH, 85540 Haar bei München

Autor/Author: Thomas Riegler
Projektmanager/Project manager: Florian Greßhake
Satz/Layout: G&U Language & Publishing Services GmbH, Flensburg
Übersetzung/Translation: G&U Language & Publishing Services GmbH, Flensburg
art & design: www.ideehoch2.de
Druck/Print: Graspo CZ, a.s.

Inhaltsverzeichnis / Table of contents

Faszinierende Autos und ihre Motoren aus 14 Jahrzehnten

Die Geschichte des Verbrennungsmotors, aber auch die des Automobils, geht zurück bis in die zweite Hälfte des 19. Jahrhunderts. Findige Entwickler wie Nicolaus August Otto und Rudolf Diesel haben die heute noch nach ihnen benannten Verbrennungsmotoren konstruiert. Sie waren es, die die Basisarbeit geleistet und erste lauffähige Maschinen gebaut haben. Dann gab es Visionäre wie Gottlieb Daimler, die das Potenzial der Verbrennungsmotoren erkannten. Sie schufen die Grundlagen für die von uns heute gelebte Mobilität, indem sie begannen, Motoren in kutschenähnliche Gefährte einzubauen, die von ganz alleine, ohne ein Pferd vorspannen zu müssen, fuhren.

Nachdem die ersten fahrbaren Untersätze und Verbrennungsmotoren vereint wurden, erkannte man rasch Verbesserungspotenziale. So hat das Auto entscheidend zur Weiterentwicklung der Motoren beigetragen. Genauso, wie der Motor seinen nicht zu vernachlässigenden Beitrag zum Automobil geleistet hat. Diese sich gegenseitig befruchtende Weiterentwicklung hält bis heute an.

Kaum eine andere Erfindung als der Verbrennungsmotor hat das 20. Jahrhundert derart entscheidend beeinflusst. Wie würde unsere Welt wohl ohne ihn aussehen? Hätten wir ein so dichtes Netz gut ausgebauter, befestigter Straßen? Gäbe es überhaupt Autobahnen? Höchstwahrscheinlich nicht, wurden sie doch erst notwendig, als Autos lernten, richtig schnell zu fahren. Motoren

und Autos haben aber auch unseren Horizont in ungeahnter Weise erweitert. Heute ist es für uns ganz normal, mal schnell 50 Kilometer in die nächste größere Stadt zu fahren. Genauso, wie es heute selbstverständlich ist, mit dem Auto ganze Kontinente zu bereisen. Erst der Verbrennungsmotor und das Auto haben es uns ermöglicht, ganz bequem fremde Länder zu bereisen und kennenzulernen.

Autos werden seit nunmehr 14 Jahrzehnten gebaut und haben sich seitdem zu weit mehr als bloßen Gebrauchsgegenständen entwickelt. Viele Fahrzeuge haben Kultstatus erlangt, andere haben Technik-Geschichte geschrieben. Dabei wird gerne vergessen, dass ein Auto nicht nur wohlgeformtes Blech auf vier Rädern ist, sondern dass in ihm auch ein Herz in Form eines Motors schlägt. Ohne ihn wäre es nutz- und wertlos. Deshalb wollen wir in diesem Buch nicht nur Autos vorstellen, sondern ganz besonders auch ihre Motoren, die sie schließlich erst zu dem machen, was sie sind.

Wir wollen aber nicht nur einen Blick in die Vergangenheit und Gegenwart wagen, sondern auch sehen, was uns die Zukunft bringt. Deuten wir die Zeichen der Zeit richtig, ist das Ende des Verbrennungsmotors in den Autos bereits eingeläutet. Die Zukunft liegt im Elektromotor, der unsere Verbrennungsmaschinen allmählich aus unseren Autos verdrängen wird.

Fascinating Cars and Their Engines from 14 Decades

The history of the combustion engine and of the motorcar goes back to the second half of the 19th century. Resourceful inventors like Nicolaus August Otto and Rudolf Diesel developed the combustion engines that bear their names to this day. They did the fundamental work and built the first working engines. Then there were visionaries like Gottlieb Daimler who recognized the potential of combustion engines. They laid the foundation for the mobility that we enjoy today by putting engines into carriages that could drive all of their own without the need to harness a horse.

After combining vehicles and combustion engines for the first time, the inventors soon realized many opportunities for improvement. In this way, the motorcar contributed significantly to the further development of engines. Similarly, the engine made an essential contribution to the development of the motorcar. This cross-fertilization lasts up to this day.

Nearly no other invention has affected the 20th century as significantly as the combustion engine. How would the world look like without it? Would we have such a dense network of wide and well-constructed streets? Would there be motorways? Probably not, as those were only required when the cars learned to drive really fast. Engines and cars have widened our horizons in unexpected ways. Today, it is nothing special to drive to the next big city 30 miles away or even to travel whole continents by car. Only the combustion engine and the car have made it possible to visit foreign countries in a convenient way.

Cars are being built for 14 decades. During this time, they have developed into much more than simple commodities. Many vehicles have gained cult status, while others made technological history. However, it is often forgotten that a car is not just shapely sheet metal on wheels but also has a heart in the form of an engine. Without the engine, a car would be useless and worthless. In this book we thus do not only present cars but particularly the engines as well since they are what makes a car a real car.

We do not just have a look at the past and the present but also dare to see what the future has in store. If we correctly interpret the signs of the times, the final days for the use of combustion engines in cars have already started. The future belongs to the electric motor, which will gradually replace the combustion engine in our cars.

Der Benz Patent-Motorwagen von 1886 gilt als das erste Automobil der Welt.
The Benz Patent-Motorwagen of 1886 is considered to be the world's first automobile.

Benz Patent-Motorwagen Nummer 1   1886

Fahrzeugdaten

Hersteller:	Benz
Land:	Deutschland
Modell:	Patent-Motorwagen Nummer 1
Bauzeit:	1884–1886
Länge:	2.700 mm
Breite:	1.400 mm
Höhe:	1.450 mm
Radstand:	1.450 mm
Leergewicht:	rund 265 kg
Antriebsart:	Hinterrad
Höchstgeschwindigkeit:	6 bis 16 km/h
Verbrauch:	10 Liter/100 km

Beim ersten von Carl Benz entwickelten stationären Benzinmotor handelte es sich um einen Einzylinder-Zweitakter, der am Silvesterabend des Jahres 1879 zum ersten Mal lief. Der große geschäftliche Erfolg des Motors ermöglichte es Carl Benz, sich zunehmend seinem Traum von einem mit Benzinmotor angetriebenen Wagen zu widmen. Bereits 1884 begann er mit den Arbeiten an seinem dreirädrigen Motorwagen, den er am 29. Januar 1886 schließlich beim Deutschen Reichspatentamt anmeldete. Damit galt das Automobil offiziell als erfunden. Der Rahmen des Patent-Motorwagens war aus gebogenen und geschweißten Stahlrohren gefertigt. Um das Lenken des an den Hinterrädern angetriebenen Fahrzeugs zu erleichtern, sah Benz nur ein Vorderrad vor, das in einer ungefederten Gabel hing und über eine mit einer Kurbel verbundene Zahnstange gesteuert wurde. Das Drehmoment des Motors wurde über ein Riemengetriebe und Ketten auf die hinteren Räder übertragen.

Specifications

Manufacturer:	Benz
Country:	Germany
Model:	Patent-Motorwagen no. 1
Produced:	1884–1886
Length:	2700 mm
Width:	1400 mm
Height:	1450 mm
Wheelbase:	1450 mm
Empty weight:	approx. 265 kg
Type of drive:	Rear-wheel drive
Max. speed:	6 to 16 km/h
Fuel consumption:	10 l/100 km

The first stationary gasoline engine developed by Carl Benz was a 2-stroke 1-cylinder model. At New Year's Eve 1879, it ran for the first time. The huge commercial success enabled Carl Benz to dedicate himself more and more to his dream of developing a car powered by a gasoline engine. In 1884, he already started to work on his three-wheel motorcar, which he finally registered at the German patent office on January 29th 1886. With this patent, the automobile was officially invented. The frame of the patent motorcar was made of bent and welded steel tubes. To facilitate steering the rear-driven vehicle, Carl Benz designed the car with only one front wheel. It was fastened inside an unsprung fork and controlled by a pinion with a crank lever. The engine torque was transferred to the rear wheels by means of a belt transmission and chains.

Der Motor

Motordaten

Bauart:	Viertakt-Motor
Zylinderzahl:	1
Hubraum:	954 cm³
Bohrung:	90 mm
Hub:	150 mm
Leistung:	0,55 kW (0,75 PS) bei 400/min

Im Patent-Motorwagen Nummer 1 war ein Einzylinder-Viertaktmotor mit einem Hubraum von rund 0,95 Litern eingebaut. Er gab seine höchste Leistung von 0,55 kW (0,75 PS) bei 400 Umdrehungen pro Minute ab. Der mit 110 Kilogramm für die damaligen Verhältnisse leichte Motor besaß eine Verdampfungskühlung. Der Einlass-Gleitschieber und das Auslass-Tellerventil wurden über eine Nockenscheibe und Exzenterstangen per Stoßstange und Kipphebel betätigt. Der von Carl Benz erdachte Oberflächen-Vergaser sorgte nicht nur für die Gemischaufbereitung, sondern diente gleichzeitig als Treibstofftank mit einem Fassungsvermögen von 1,5 Litern. Als Zündkerze kam eine Eigenentwicklung zum Einsatz. Wie spätere Analysen ergaben, stimmte der dabei verwendete Elektroden-Werkstoff mit handelsüblichen Zündkerzen der 1930er-Jahre überein. Benzin im heutigen Sinne gab es für diesen Motor noch nicht. Deshalb musste Carl Benz auf ein damals in Apotheken unter dem Namen Ligroin angebotenes Alkohol-Benzingemisch zurückgreifen.

The engine

Engine specifications

Type:	4-stroke engine
Number of cylinders:	1
Displacement:	954 cm³
Bore:	90 mm
Stroke:	150 mm
Power:	0.55 kW (0.75 hp) at 400/min

The Patent-Motorwagen No. 1 had a 4-stroke 1-cylinder engine with a displacement of approx. 0.95 l. Its maximum power amounted to 0.55 kW (0.75 hp) at 400 rpm. At 110 kg, the engine was relatively light for its days. It used an evaporative cooling. The intake sliding spool and the outlet disk valve were controlled by push rods and rocker levers. The surface carburetor invented by Benz did not only prepare the mix but also served as fuel tank with a capacity of 1.5 liters. The spark plug was also an in-house development. According to later analysis, the electrode material was identical to the material used for commercially available spark plugs in the 1930s. Gasoline in the modern sense was not available for this engine. Benz had to use an alcohol-gasoline mix, which was sold as Ligroin in pharmacies.

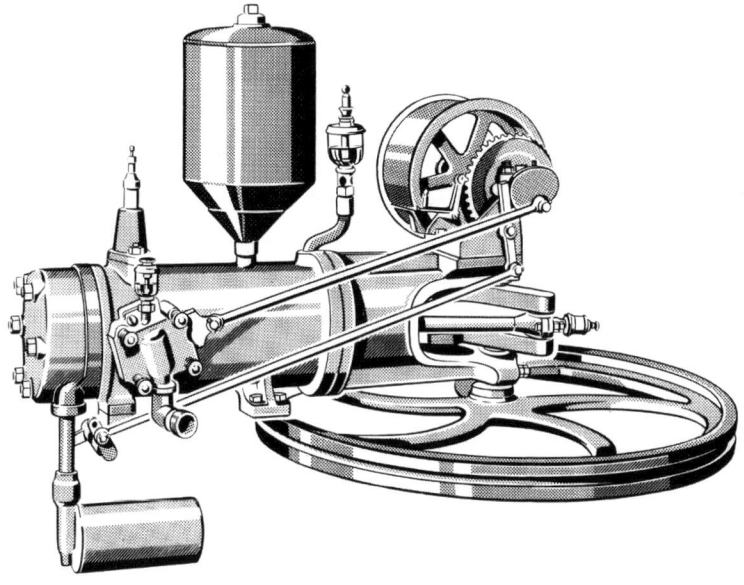

Blick von oben auf den im Heck des Benz Patent-Motorwagens angeordneten Motor.
Top view of the engine in the rear part of the Benz Patent-Motorwagen.

Grafik des im Benz Patent-Motorwagens eingebauten Einzylinder-Motors.
The 1-cylinder engine used in the Benz Patent-Motorwagen.

Mit der Patentierung des Benz Patent-Motorwagens Nummer 1 im Jahr 1886 galt das Automobil offiziell als erfunden. Mit 0,55 kW (0,75 PS) Leistung und maximal 16 km/h war es aber ein beschaulicher Beginn.

With patenting the Benz Patent-Motorwagen No. 1 in 1886, the automobile was officially invented. With its power of 0.55 kW (0.75 PS) and a top speed of 16 km/h, the beginnings were rather humble.

Der Stahlradwagen wurde 1889 auf der Pariser Weltausstellung erstmals der Öffentlichkeit präsentiert.
The steel-wheeled-car war first introduced at the Paris World's Fair in 1889.

Daimler Stahlradwagen 1889

Fahrzeugdaten

Hersteller:	Daimler
Land:	Deutschland
Modell:	Stahlradwagen
Bauzeit:	1889
Länge:	2.350 mm
Breite:	1.450 mm
Höhe:	1.450 mm
Radstand:	1.400 mm
Leergewicht:	268 kg
Antriebsart:	Hinterrad
Höchstgeschwindigkeit:	18 km/h
Verbrauch:	k.A.

Der von Gottlieb Daimler und Wilhelm Maybach entwickelte Stahlradwagen stellte eine eigenständige, ganzheitliche Konstruktion dar und ist als echtes Automobil zu bezeichnen. Ältere Autos waren im Wesentlichen umgebaute Kutschen, auch Motorkutschen genannt. 1889 wurde der Stahlradwagen auf der Pariser Weltausstellung vorgestellt. In ihm wurden erstmals ein Zweizylinder-Motor und ein Viergang-Zahnradgetriebe umgesetzt. Zur Realisierung der Lenkung wurden die beiden Vorderräder nach Fahrradart einzeln in Gabeln geführt und durch eine Spurstange verbunden. Diese Neuerungen gaben den Anstoß für die Entstehung der Automobilindustrie in Frankreich. Die ersten 1890 von Peugeot und 1891 von Panhard & Levassor gebauten Autos basierten auf dem Stahlradwagen und wurden von in Lizenz produzierten Daimler-Motoren angetrieben. Eine modifizierte Version des Stahlradwagens wurde von 1892 bis 1895 von der DMG, der Daimler-Motoren-Gesellschaft, als Daimler Motorwagen vertrieben.

Specifications

Manufacturer:	Daimler
Country:	Germany
Model:	Stahlradwagen
Produced:	1889
Length:	2350 mm
Width:	1450 mm
Height:	1450 mm
Wheelbase:	1400 mm
Empty weight:	268 kg
Type of drive:	Rear-wheel drive
Max. speed:	18 km/h
Fuel consumption:	n.a.

The steel-wheeled car developed by Gottlieb Daimler and Wilhelm Maybach was an original, integral construction and thus a real automobile. Older cars were more or less modified stagecoaches and called "motor coaches." The steel-wheeled car was introduced at the Paris World's Fair in 1889. It was the first car to use a 2-cylinder engine and a 4-gear transmission. Steering was enabled by guiding both front wheels in forks (similar to bicycle wheels) and connecting them by a steering link. These innovations kicked off the formation of the French automobile industry. The first cars built by Peugeot in 1890 and by Panhard & Levassor in 1891 based on the steel-wheeled car and were driven by license-built Daimler engines. A modified version of the steel-wheeled car was sold as Daimler Motorwagen ("Daimler motor car") by the Daimler-Motoren-Gesellschaft (DMG) from 1892 to 1895.

Der Motor

Motordaten

Bauart:	Viertakt-17-Grad-V-Motor
Zylinderzahl:	2
Hubraum:	565 cm³
Bohrung:	60 mm
Hub:	100 mm
Leistung:	1,1 kW (1,5 PS)
	bei 700/min

Der von Daimler und Maybach 1889 gefertigte Zweizylinder-Motor basierte auf dem 1885 gebauten Einzylinder-Motor, der wegen seiner Bauhöhe auch als sogenannte Standuhr in die Geschichte einging. Die „Standuhr" war der weltweit erste kleine schnell laufende Benzin-Verbrennungsmotor, der leicht und stark genug war, ein Fahrzeug anzutreiben. Dieser 462 cm³ große Einzylinder-Motor brachte es auf eine Leistung von 0,8 kW (1,1 PS) bei 650 Umdrehungen pro Minute. Beim ersten Zweizylinder-Motor waren die beiden Zylinder V-förmig mit einem Winkel von 17 Grad zueinander angeordnet. Bei einer Drehzahl von 700 Umdrehungen pro Minute entfaltete er seine maximale Leistung von 1,1 kW (1,5 PS). Dieser Viertaktmotor besaß ein hängendes, automatisch arbeitendes Einlassventil sowie ein seitlich stehendes gesteuertes Auslassventil. Als Ventilsteuerung kam eine Kurvenuntersteuerung zum Einsatz. Ein Oberflächenvergaser mit integriertem Kraftstoffvorrat übernahm die Gemischbildung.

The engine

Engine specifications

Type:	4-stroke V2 engine (17°)
Number of cylinders:	2
Displacement:	565 cm³
Bore:	60 mm
Stroke:	100 mm
Power:	1,1 kW (1,5 hp)
	at 700/min

The 2-cylinder engine built by Daimler and Maybach in 1889 based on the 1-cylinder engine of the 1885 built 1-cylinder motor, which was nicknamed "grandfather clock" ("Standuhr") because of its height. The "grandfather clock" was the first small and fast running gasoline combustion engine in the world that was light and powerful enough to drive a vehicle. With its displacement of 462 cm³, the engine provided 0.8 kW (1.1 hp) at 650 rpm. The two cylinders of the first 2-cylinder engine were mounted in V-arrangement at an angle of 17°. At 700 rpm, the engine showed its maximum power of 1.1 kW (1.5 hp). This 4-stroke engine had an automatic overhead intake valve and a controlled outlet valve at the side. Valve control was implemented by curve understeering. A surface carburetor with integrated fuel supply was used to provide the mixture.

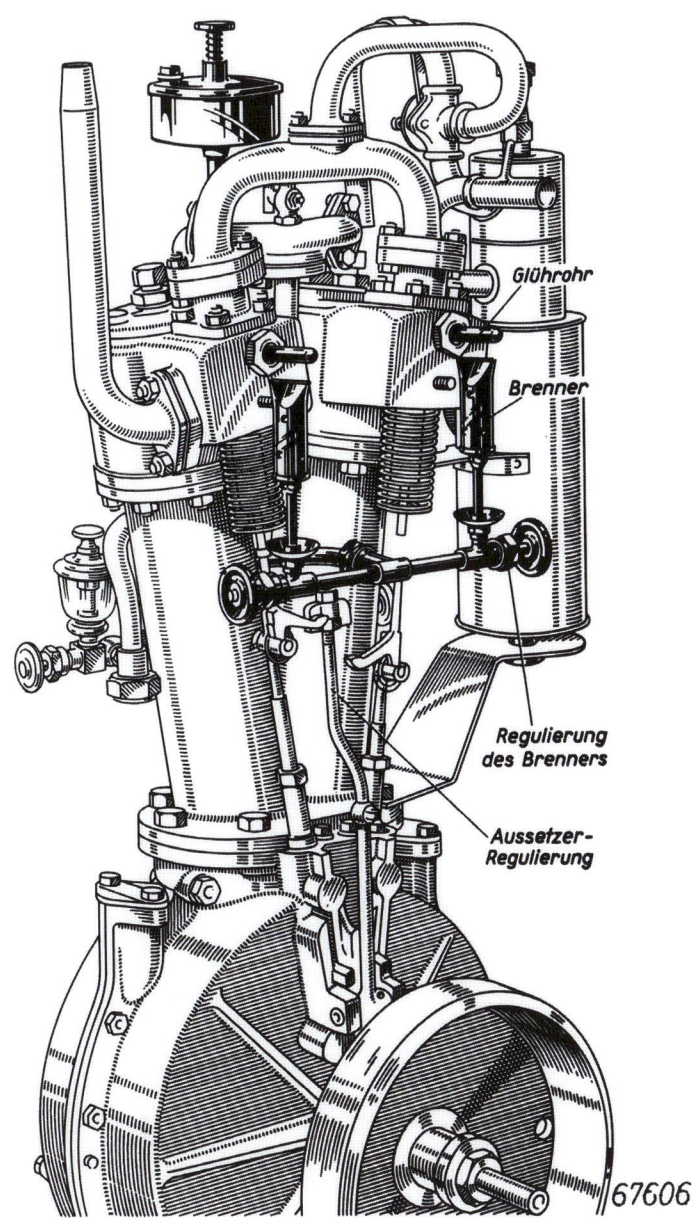

Glührohr

Brenner

Regulierung
des Brenners

Aussetzer-
Regulierung

67606

Der 17-Grad-V2-Motor des Daimler-Stahlradwagens leistete 1,1 kW (1,5 PS).
The 17° V2 engine of the Daimler steel-wheeled car provided 1.1 kW (1.5 hp).

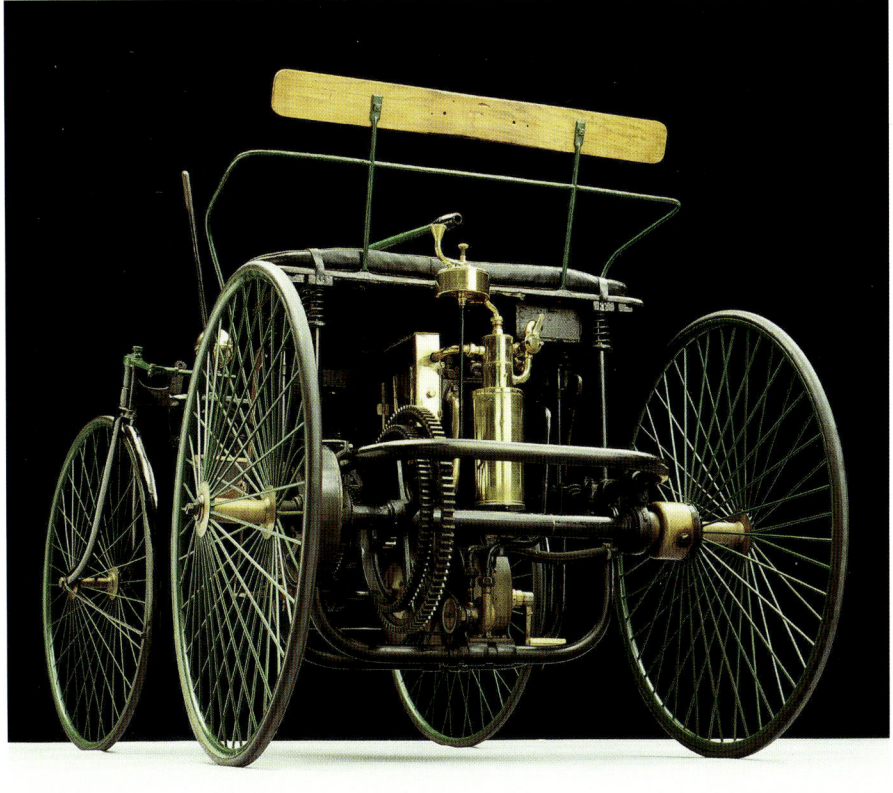

In eingebautem Zustand ist die Höhe des V2-Viertakt-Motors gut zu erkennen.
The height of the 4-stroke V2 engine can be easily recognized when mounted.

Im Stahlradwagen von 1889 setzten Gottlieb Daimler und Wilhelm Maybach erstmals einen Zweizylinder-Motor in Verbindung mit einem Viergang-Zahnrad-getriebe ein.

In the steel-wheeled car of 1889, Gottlieb Daimler and Wilhelm Maybach used a combination of a 2-cylinder engine and a 4-gear transmission for the first time.

Im Benz Dos-à-Dos von 1899 saßen Fahrer,
Beifahrer und Passagiere Rücken an Rücken.

In the Benz Dos-à-Dos of 1899, driver and
passengers sat back to back.

U 16545

Benz Dos-à-Dos

Fahrzeugdaten

Hersteller:	Benz
Land:	Deutschland
Modell:	Dos-à-Dos
Bauzeit:	1899–1901
Länge:	3.300 mm
Breite:	1.850 mm
Höhe:	2.200 mm
Radstand:	k.A.
Leergewicht:	650–1.300 kg
Antriebsart:	Hinterrad
Höchstgeschwindigkeit:	35–40 km/h
Verbrauch:	k.A.

Beim Benz Dos-à-Dos handelt es sich um den Nachfolger des Benz Patent-Motorwagens Victoria. Er wurde von 1899 bis 1901 in mehreren viersitzigen Karosserieversionen gebaut. Den Ausführungen gemeinsam waren zwei Sitzbänke für je zwei Personen, auf denen die Fahrgäste Rücken an Rücken (Dos-à-Dos) Platz nahmen. Der Wagen war mit Holzspeichenrädern ausgestattet, auf die wahlweise Vollgummi- oder Luftreifen aufgezogen waren. Die Radaufhängungen vorne und hinten übernahmen Starrachsen mit Vollelliptik-Blattfedern. Der Dos-à-Dos besaß ein dreistufiges Vorgelegegetriebe mit Rückwärtsgang, das über Ketten auf die beiden Hinterräder wirkte. Im Dos-à-Dos-Wagen kam als Antriebsquelle ein Viertakt-Boxermotor mit zwei Zylindern zum Einsatz. Dieser wurde seiner gegenläufigen Kolben wegen von Carl Benz damals noch als Contra-Motor bezeichnet. Da dieser Motor von Benz laufend verbessert wurde, waren für den Dos-à-Dos von 1899 bis 1901 mehrere Motor-Versionen verfügbar.

Specifications

Manufacturer:	Benz
Country:	Germany
Model:	Dos-à-Dos
Produced:	1899–1901
Length:	3300 mm
Width:	1850 mm
Height:	2200 mm
Wheelbase:	n.a.
Empty weight:	650–1300 kg
Type of drive:	Rear-wheel drive
Max. speed:	35–40 km/h
Fuel consumption:	n.a.

The Benz Dos-à-Dos was the successor of the Benz Patent-Motorwagen Victoria. It was built with various four-seater auto bodies from 1899 to 1901. All versions had two benches for two people each, where the passengers sat down back to back ("dos-à-dos"). The wheels of the car had wooden spokes and could be equipped with solid tires or pneumatic tires. Front and rear wheel were suspended on fixed axles which were equipped with fully elliptic leaf springs. The Dos-à-Dos had a three-step auxiliary shaft transmission with reverse gear, which affected the rear wheels via chains. The car was driven by a four-stroke flat 2-cylinder engine. Because of its opposing pistons, this engine concept was called "contra motor" by Carl Benz. As Benz continuously improved the engine, several engine versions for the Dos-à-Dos where available from 1899 to 1901.

Der Motor

Motordaten

Bauart:	Viertakt-Boxermotor
Zylinderzahl:	2
Hubraum:	2.690 cm³
Bohrung:	120 mm
Hub:	120 mm
Leistung:	5,9 kW (8 PS)
	bei 920/min

Im Jahr 1896 schlug die Geburtsstunde des Viertakt-Boxermotors, der von Carl Benz entwickelt worden war. Bei ihm wirkten zwei gegenläufig arbeitende Kolben in sich gegenüberstehenden Zylindern auf eine gemeinsame Kurbelwelle. Bis 1899 wurde der Contra-Motor zur Serienreife entwickelt. Bekannt wurde der Contra-Motor, als er in einer 1,7-Liter-Variante mit 3,7 kW (5 PS) bei 940 Touren in den Dos-à-Dos eingebaut wurde. Während dieser kleine Motor bis 1900 im Programm blieb, gab es auch eine größere 2,7-Liter-Version, die ständig verbessert wurde. In der Variante von 1899 leistete er 5,9 kW (8 PS). Das Modell von 1900 brachte es bereits auf 6,6 kW (9 PS) und jenes von 1901 sogar auf 7,4 kW (10 PS). Die maximale Leistung wurde jeweils bei 920 Touren abgegeben. 1902 wurde die Drehzahl des Motors auf 980 Umdrehungen pro Minute gesteigert, sodass ihm 8,8 kW (12 PS) entlockt werden konnten. Schließlich wurde auch der 1,7-Liter-Motor auf 6,6 kW (9 PS) verbessert.

The engine

Engine specifications

Type:	Four-stroke flat engine
Number of cylinders:	2
Displacement:	2690 cm³
Bore:	120 mm
Stroke:	120 mm
Power:	5.9 kW (8 hp)
	at 920 rpm

Developed by Carl Benz, the four-stroke flat engine was born in 1896. In two cylinders facing each other, two pistons working in opposite directions affect a common crankshaft. The "contra motor" was developed to production readiness until 1899. At this time, a 1.7-liter version with 3.7 kW (5 hp) at 940 rpm was mounted in the Dos-à-Dos. While this small engine remained in the portfolio until 1900, there was also a larger 2.7-liter version, which was continually improved. The 1899 model delivered 5.9 kW (8 hp), the 1900 model 6.6 kW (9 hp) and the 1901 model even 7.4 kW (10 hp). Maximum power was reached at 920 rpm. In 1902, the engine speed was increased to 980 rpm, which now provided 8.8 kW (12 hp). Eventually, the 1.7-liter engine was improved to yield a power of 6.6 kW (9 hp).

Der 1896 von Carl Benz entwickelte Contra-Motor gilt als Urvater der Boxermotoren.
The contra motor developed by Carl Benz in 1896 is considered the forefather of the flat engine.

Der Benz Dos-à-Dos von 1889 ist nicht nur für die Sitzanordnung der Fahr-gäste Rücken an Rücken bekannt, sondern auch für den von Carl Benz ent-worfenen Viertakt-Boxermotor.

The 1899 Benz Dos-à-Dos is not only remarkable for the back-to-back seating but also for the four-stroke flat engine developed by Carl Benz.

Namensgebung: Der erste Mercedes wurde nach der Tochter von Auftraggeber Jellinek benannt.

Naming: The first Mercedes was named after the daughter of its initiator Jellinek.

Mercedes 35 PS

Fahrzeugdaten

Hersteller:	Daimler
Land:	Deutschland
Modell:	Mercedes 35 PS
Bauzeit:	1900–1902
Länge:	k.A.
Breite:	1.345 mm
Höhe:	k.A.
Radstand:	2.345 mm
Leergewicht:	1.000 kg
Antriebsart:	Hinterrad
Höchstgeschwindigkeit:	90 km/h
Verbrauch:	k.A.

Im Jahr 1900 entwickelte Konstrukteur Wilhelm Maybach für die Daimler-Motoren-Gesellschaft den Mercedes 35 PS. Er war nach heutigen Maßstäben das erste moderne Automobil, das deutlich vom zuvor vorherrschenden Kutschenbauprinzip abwich. Mit Innovationen wie dem Bienenwabenkühler läutete dieses Fahrzeug eine technische Revolution ein – und wurde zum Maßstab für den Automobilbau. Der für damalige Verhältnisse lange Radstand und die breite Spur bildeten die Grundlage für ein stabiles Fahrverhalten. Das erste Exemplar dieses Modells wurde am 22. Dezember 1900 an den Auftraggeber Emil Jellinek ausgeliefert. Der erfolgreiche Geschäftsmann Jellinek hatte die Neukonstruktion nach seiner Tochter Mercedes benannt. Die Modellbezeichnung für alle künftigen Fahrzeuge der Daimler-Motoren-Gesellschaft war geboren. Im Frühjahr 1901 feierte der Mercedes 35 PS spektakuläre Rennerfolge – was die Presse veranlasste vom „Beginn der Ära Mercedes" zu sprechen.

Specifications

Manufacturer:	Daimler
Country:	Germany
Model:	Mercedes 35 PS
Produced:	1900–1902
Length:	n.a.
Width:	1345 mm
Height:	n.a.
Wheelbase:	2345 mm
Empty weight:	1000 kg
Type of drive:	Rear-wheel drive
Max. speed:	90 km/h
Fuel consumption:	n.a.

In 1900, constructing engineer Wilhelm Maybach developed the Mercedes 35 PS for the Daimler-Motoren-Gesellschaft ("Daimler motor company"). This was the first modern automobile in the present day sense, as it significantly deviated from the previously dominating stagecoach-type construction principle. With innovations like the honeycomb radiator, this vehicle heralded a technological revolution and become a benchmark for automobile construction. The long wheelbase for its time and the wide track enabled a steady ride. The first copy of this model was shipped to its initiator, Emil Jellinek, on December 22nd 1900. The successful businessman had named the new model after his daughter Mercedes. Thus the model name for all future vehicles of the Daimler-Motoren-Gesellschaft was born. In spring 1901, the Mercedes 35 PS gained spectacular racing wins, which caused the press to proclaim the "beginning of the Mercedes era."

Der Motor

Motordaten

Bauart:	Viertakt-Reihenmotor
Zylinderzahl:	4
Hubraum:	5.913 cm³
Bohrung:	116 mm
Hub:	140 mm
Leistung:	26 kW (35 PS)
	bei 950/min

Der im Mercedes 35 PS eingebaute Vierzylinder-Reihenmotor war rund 230 Kilogramm schwer – und damit rund 90 Kilogramm leichter als das Triebwerk des Vorgängermodells Phönix. Der Motor war vorne eingebaut und mit dem erstmals aus Stahlblech gepressten Rahmen verschraubt. Der Vierzylinder brachte es auf – für damalige Verhältnisse – sensationelle 26 kW (35 PS) bei 950 Umdrehungen pro Minute. Das Kurbelgehäuse des Motors wurde erstmals aus Aluminium gefertigt. Weiter bildeten Zylinder und Zylinderkopf eine Einheit. Neu war zudem, dass die Einlassventile über eine Nockenwelle gesteuert wurden. Pro Zylinderpaar war je ein Vergaser vorgesehen. Über einen Hebel am Lenkrad ließ sich die Drehzahl zwischen 300 und 1.000 Umdrehungen pro Minute regeln. Zur Ausstattung des Motors gehörte eine Niederspannungs-Magnetzündung. Das Getriebe des Mercedes 35 PS verfügte über vier Vorwärtsgänge und einen Rückwärtsgang. Der Motor trieb die Hinterräder des Mercedes an.

The engine

Engine specifications

Type:	Four-stroke in-line engine
Number of cylinders:	4
Displacement:	5913 cm³
Bore:	116 mm
Stroke:	140 mm
Power:	26 kW (35 hp)
	at 950/min

The 4-cylinder in-line engine used in the Mercedes 35 PS weighed approx. 230 kg and thus 90 kg less than the engine of the predecessor model, Phönix. The engine was mounted in the front and bolted to the frame, which was the first frame made from pressed sheet-steel. The engine provided 26 kW (35 hp) at 950 rpm, which was a tremendous power for its time. The crankcase of the engine was made from aluminum for the first time. The cylinder and the cylinder head were now one unit. As another new feature, the intake valves were now controlled by a camshaft. One carburetor was used per cylinder pair. A lever at the steering wheel allowed regulating the engine speed in a range between 300 and 1000 rpm. The engine was equipped with a low-voltage magneto ignition. The transmission of the Mercedes 35 PS had four forward gears and one reverse gear. The engine drove the rear wheels of the car.

Der 5,9-Liter-Vierzylinder-Reihenmotor des Mercedes besaß ein Kurbelgehäuse aus Aluminium.

The crankcase of the 5.9-liter 4-cylinder in-line engine of the Mercedes was made of aluminum.

Der Vierzylinder des Mercedes 35 PS wog volle 90 Kilogramm weniger als der Motor des Vorgängermodells Phönix.

The 4-cylinder engine of the Mercedes 35 PS weighed 90 kg less than the engine of the predecessor model Phönix.

Daimler-Konstrukteur Wilhelm Maybach entwickelte mit dem Mercedes 35 PS ein Modell, das deutlich vom zuvor üblichen Prinzip der motorisierten Kutschenwagen abwich. Er schuf damit das erste moderne Automobil.

With the Mercedes 35 PS, Daimler engineer Wilhelm Maybach developed a model that significantly deviated from the previously dominating principle of a motorized coach, and thus created the first modern car.

Es existieren drei Exemplare des legendären Benz 200 PS.
Der hier abgebildete Wagen steht im Mercedes-Benz-Museum.

Three specimen of the legendary Benz 200 PS are still in existence.
This one is on exhibit in the Mercedes-Benz Museum.

Benz 200 PS „Blitzen-Benz"   1909

Benz 200 PS „Blitzen-Benz"

Fahrzeugdaten

Hersteller:	Benz
Land:	Deutschland
Modell:	200 PS, auch Blitzen-Benz
Bauzeit:	1909
Länge:	4.820 mm
Breite:	1.600 mm
Höhe:	1.280 mm
Radstand:	2.800 mm
Leergewicht:	1.450 kg
Antriebsart:	Hinterrad
Höchstgeschwindigkeit:	228 km/h
Verbrauch:	k.A.

Mit dem Benz 200 PS entstand 1909 ein faszinierendes Auto, das Rekorde brechen sollte: Die Grenze von 200 km/h war das Ziel; dafür bauten die Techniker bei Benz einen Vierzylinder-Motor mit gigantischen 21,5 Liter Hubraum. 147 kW (200 PS) leistete dieses 407 Kilogramm schwere Triebwerk. Der Franzose Victor Hémery fuhr mit dem Benz im November 1909 im britischen Brooklands 202,7 km/h. Erstmals war damit die 200-km/h-Grenze in Europa durchbrochen worden. Die Rekordjagd ging für den bald „Blitzen-Benz" genannten Boliden in den USA weiter: Barney Oldfield erreichte 1910 in Daytona Beach in Florida 211,97 km/h. Doch die Bestmarke wurde nicht als neuer Weltrekord anerkannt. Die große Stunde des Benz 200 PS schlug letztlich am 23. April 1911: Es war der US-Amerikaner Bob Burman, der auf dem Daytona Beach alle bisherigen Geschwindigkeitsrekorde überbot. Mit 228,1 km/h war der Benz dabei schneller als jedes andere Auto, Schienenfahrzeug und Flugzeug zuvor.

Specifications

Manufacturer:	Benz
Country:	Germany
Model:	200 PS, also called Blitzen Benz
Produced:	1909
Length:	4820 mm
Width:	1600 mm
Height:	1280 mm
Wheelbase:	2800 mm
Empty weight:	1450 kg
Type of drive:	Rear-wheel drive
Max. speed:	228 km/h
Fuel consumption:	n.a.

In 1909, the Benz 200 PS was created – a fascinating car that was designed to break records. The objective was to reach the limit of 200 km/h. To this end, the Benz technicians built a 4-cylinder engine with an enormous displacement of 21.5 liters. The engine weighed 407 kg and provided 137 kW (200 hp). In November 1909, the Frenchman Victor Hémery drove the Benz in the Brooklands aerodrome in Britain at 202.7 km/h. Thus the 200 km/h limit was exceeded for the first time in Europe. The hunt for records continued with the car soon dubbed "Blitzen Benz" continued in the USA. In 1910, Barney Oldfield reached a top speed of 211.97 km/h in Daytona Beach, Florida. However, this top mark was not recognized as a new world record. The big moment of the Benz 200 PS finally came on April 23rd 1911, when the American Bob Burman surpassed all previous speed records. At 228.1 km/h, the Benz was faster than any car, rail vehicle or aircraft before.

Der Motor

Motordaten

Bauart:	Viertakt-Reihenmotor
Zylinderzahl:	4
Hubraum:	21.504 cm³
Bohrung:	185 mm
Hub:	200 mm
Leistung:	147 kW (200 PS) bei 1.600/min

Der Motor des „Blitzen-Benz" basierte auf dem Triebwerk des Benz Grand-Prix-Rennwagens von 1908. Dieser Rennmotor leistete ursprünglich 110 kW (150 PS). Zur Leistungssteigerung wurde die Bohrung des Vierzylinders im „Blitzen-Benz" auf 185 Millimeter vergrößert, sodass der Hubraum von 15,1 auf unglaubliche 21,5 Liter stieg. Er war damit zugleich der hubraumstärkste Motor, den Benz je in einen Rennwagen einbaute. In seiner ersten Variante brachte er es auf 135 kW (184 PS), später stieg die Leistung auf 147 kW (200 PS) bei 1.600/min. Der Reihenmotor des Benz war mit je zwei hängenden Ventilen pro Zylinder und einer seitlichen Nockenwelle ausgestattet, die über Zahnräder angetrieben wurde. Die Kurbelwelle war fünffach gelagert. Es war übrigens echte Handarbeit, den Motor des „Blitzen-Benz" während der Fahrt mit dem benötigten Treibstoff zu versorgen: Dazu musste der Beifahrer eine Druckluft-Handpumpe betätigen, mit der das Benzin vom Tank zum Motor befördert wurde.

The engine

Engine specifications

Type:	Four-stroke in-line engine
Number of cylinders:	4
Displacement:	21,504 cm³
Bore:	185 mm
Stroke:	200 mm
Power:	147 kW (200 hp) at 1600/min

The engine of the "Blitzen Benz" was based on the engine of the Benz Grand Prix racing car of 1908. Originally, this racing engine provided 110 kW (150 hp). To improve the power, the bore of the 4-cylinder engine of the "Blitzen Benz" was extended to 185 mm, thus increasing the displacement of 15.1 liters to an incredible 21.5 liters. This was the engine with the largest displacement that Benz ever mounted in a racing car. The first variant achieved 135 kW (184 hp), which was increased to 147 kW (200 hp) at 1600 rpm in later models. The in-line engine had two overhead valves per cylinder and a gear-driven camshaft at the side. The crankshaft was supported at five points. Incidentally, the engine of the "Blitzen Benz" had to be manually supplied with fuel during the ride: The co-driver had to operate a pneumatic hand pump to transport gasoline from the tank to the engine.

Dieser Vierzylinder war der hubraumstärkste Motor, den Benz je in einem Rennwagen einsetzte.

The 4-cylinder engine had the largest displacement of any engines that Benz ever mounted in racing cars.

Das Triebwerk basierte auf jener Konstruktion, die Benz im Grand-Prix-Rennwagen von 1908 genutzt hatte.

The engine was based on the design that Benz used in the Grand Prix racing car of 1908.

Der Benz 200 PS besaß einen gigantischen 21,5-Liter-Motor. Dieser Vierzylinder leistete 147 kW (200 PS) bei 1.600/min – und beschleunigte den „Blitzen Benz" 1911 bei seiner Weltrekordfahrt auf 228,1 km/h.

The Benz 200 PS had a giant 21.5-liter 4-cylinder engine, which provided 147 kW (200 hp) at 1600 rpm and accelerated the "Blitzen Benz" to 228,1 km/h on its world record ride in 1911.

Der „Alpensieger" auf großer Bergfahrt – rund 100 Jahre, nachdem er seine großen Siege feierte.

The "Alpensieger" in the mountains approximately 100 years after its famous wins.

August Horch persönlich lenkte den Audi Typ C 14/35 PS auf der Österreichischen Alpenfahrt 1914.

August Horch himself at the wheel of the Audi Type C 14/35 PS during the Austrian Alpine Rally in 1914.

Sieger August Horch mit dem Audi Typ C 14/35 PS bei der Österreichischen Alpenfahrt.

Winner August Horch with an Audi Type C 14/35 PS at the Austrian Alpine Rally.

Audi Typ C 14/35 PS „Alpensieger"   1911

Fahrzeugdaten

Hersteller:	Audi
Land:	Deutschland
Modell:	Typ C 14/35 PS
Baujahr:	1911–1925
Länge:	4.700 mm
Breite:	1.780 mm
Höhe:	2.000 mm
Radstand:	2.896 mm (variabel)
Leergewicht:	1.750 kg
Antriebsart:	Hinterrad
Höchstgeschwindigkeit:	90–100 km/h
Verbrauch:	17 Liter/100 km

Der dritte von August Horch für die Firma Audi entwickelte Wagen wurde intern als „Typ C" bezeichnet. Offiziell trug er jedoch die Modellbezeichnung 14/35 PS, was für 14 Steuer- und 35 Motor-PS stand. Der Typ C 14/35 PS wurde 1911 vorgestellt und zählte zu Horchs besten und ausgereiftesten Konstruktionen. Zugleich war der Wagen das erste motorsportliche Fahrzeug der Marke Audi. Er beendete die Österreichische Alpenfahrt – seinerzeit eine der schwersten und anspruchsvollsten Langstreckenfahrten überhaupt – in den Jahren 1912 bis 1914 dreimal hintereinander als Sieger. Doch der Typ C war keineswegs ein spezialisierter Rennwagen, sondern ein sportlich ausgelegter Tourenwagen mit 35 PS Leistung und 90 bis 100 km/h Höchstgeschwindigkeit. Vom Typ C wurden 1.116 Einheiten gebaut, darunter auch einige Ausführungen für das Militär. Sogar eine Lkw-Variante gab es von diesem Wagen. Sie war bis 1928 erhältlich und wurde mehr als 300 Mal hergestellt.

Specifications

Manufacturer:	Audi
Country:	Germany
Model:	Type C 14/35 PS
Produced:	1911–1925
Length:	4700 mm
Width:	1780 mm
Height:	2000 mm
Wheelbase:	2896 mm (variable)
Empty weight:	1750 kg
Type of drive:	Rear-wheel drive
Max. speed:	90–100 km/h
Fuel consumption:	17 l/100 km

The third car that August Horch developed for the Audi company was internally called "type C." Its official model designation, however, was 14/35, which stood for 14 hp (in German: PS) steering power and 35 hp engine power. The Type C 14/35 PS was introduced in 1911 and is one of Horch's best and most mature designs. The car was also the first motorsports vehicle of the Audi brand. In the years 1912 to 1914, it completed the Austrian Alpine Rally three times consecutively as a winner. At that time, this was one of the most difficult and challenging long-range rallies. But the Type C was by no means a dedicated racing car but a sporty touring car with an engine power of 35 hp and a top speed of 90 to 100 km/h. 1116 units of the Type C were built, including some military versions. There was even a motor truck version of this car. It was available until 1928. More than 300 units were built.

Der Motor

Motordaten

Bauart:	Viertakt-Reihenmotor
Zylinderzahl:	4
Hubraum:	3.563 cm³
Bohrung:	90 mm
Hub:	140 mm
Leistung:	26,1 kW (35,5 PS) bei 1.800/min

Im Audi Typ C 14/35 PS „Alpensieger" sorgte ein 3,6-Liter-Saugmotor mit vier Zylindern in Reihenanordnung für Vortrieb. Bei moderaten 1.800 Umdrehungen pro Minute entwickelte der 3.563 cm3 große Motor mit dem aus heutiger Sicht bemerkenswert langen Hub von 140 Millimetern eine maximale Leistung von 26,1 kW (35,5 PS). Pro Zylinder verfügte er über zwei Ventile und eine sogenannte IOE-Steuerung (Inlet over Exhaust), die auch als Gegensteuerung bezeichnet wird. Dabei wurden die Ein- und Auslassventile von einer unten liegenden Nockenwelle gesteuert. Die oben hängenden Einlassventile wurden wie bei einem OHV-gesteuerten Motor (Overhead Valves) über Stoßstangen und Kipphebel bewegt und öffneten nach unten, während die darunter angeordneten Auslassventile nach oben und damit entgegen den Einlassventilen öffneten. Sie wurden direkt von der Nockenwelle bewegt. Die Kraftübertragung an die Hinterräder erfolgte über ein Viergang-Vorgelege-Getriebe und eine Kardanwelle.

The engine

Engine specifications

Type:	Four-stroke in-line engine
Number of cylinders:	4
Displacement:	3563 cm³
Bore:	90 mm
Stroke:	140 mm
Power:	26,1 kW (35,5 hp) at 1.800/min

The Audi Type C 14/35 PS "Alpensieger" was driven by a 3.6-liter suction engine with four cylinders in line. At moderate speeds of 1800 rpm, the 3563 cm³ engine used a stroke of 140 mm, which today seems extraordinarily long, to produce a maximum power of 26.1 kW (35.5 hp). It was equipped with two valves per cylinder and an inlet-over-exhaust (IOE) control, which is also called F-head. Here, the inlet and exhaust valves are controlled by a camshaft below. As in an OHV (Overhead valves) engine, the inlet valves above are controlled by push rods and rocker levers and open downwards, while the exhaust valves below are controlled directly by the camshaft and open upwards. Power transmission to the rear wheels occurs via a four-gear auxiliary shaft gear and a cardan shaft.

14/35 PS. Original-Audi-Chassis 4 Zylinder

für alle Arten Karosserien

mit 4 Geschwindigkeiten und Rücklauf, mit vorderen Kotflügeln,
inklusive Zubehörteile und Werkzeuge.

Steuerbetrag: 170 Mark.   Steuerbetrag: 170 Mark.

Audi

Preis des Chassis mit magnetelektrischer Lichtbogen-Zündung
===== **10500 Mark.** =====

Dieses Inserat pries den Typ C 14/35 PS
als Chassis mit Antrieb für verschiedene
Karosserien an.

This ad praised the Type C 14/35 PS as a
driven chassis for various auto bodies.

Der Motor des Audi Typ C 14/35 PS leistete
26,1 kW (35,5 PS) bei 1.800 Touren.

The engine of the Audi Type C 14/35 PS
provided 26.1 kW (35,5 hp) at 1800 rpm.

Die Erfolge bei der Österreichischen Alpenfahrt gaben dem Audi Typ C
14/35 PS den Beinamen „Alpensieger". Von 1912 bis 1914 gewann er den
Wettbewerb dreimal in Folge.

Due to its successes in the Austrian Alpine Rally, the Audi Type C 14/35 PS
earned the nickname "Alpensieger" ("Alpine winner"). The car won the Rally
three times in a row from 1912 to 1914.

Berühmte Historie: Das originale Fahrzeug von Grand-Prix-Sieger Christian Lautenschlager.
A famous piece of history: the original vehicle of Grand Prix winner Christian Lautenschlager.

Mercedes-Benz 18/100 Grand-Prix-Rennwagen

Fahrzeugdaten

Hersteller:	Daimler
Land:	Deutschland
Modell:	Mercedes 18/100 Grand-Prix-Rennwagen
Bauzeit:	1914
Länge:	4.100 mm
Breite:	1.700 mm
Höhe:	1.400 mm
Radstand:	2.845 mm
Leergewicht:	1.082 kg
Antriebsart:	Hinterrad
Höchstgeschwindigkeit:	180 km/h
Verbrauch:	k.A.

Die Geschichte der Grand-Prix-Rennen begann in Frankreich: Dort fand 1906 der erste Automobil-Wettbewerb mit der Bezeichnung „Grand Prix" auf einer Rundstrecke statt – organisiert vom Automobile Club de France. Der französische Grand Prix wurde schnell zum Saisonhöhepunkt des internationalen Motorsports. 1908 feierte der deutsche Pilot Christian Lautenschlager auf dem Kurs von Dieppe einen spektakulären Sieg – und bescherte Mercedes den ersten Grand-Prix-Erfolg. Erst 1914 trat das Mercedes-Team wieder beim Großen Preis von Frankreich an. Die Veranstalter hatten im Herbst 1913 für das Rennen ein neues technisches Reglement erlassen, sodass die Ingenieure um Paul Daimler in nur wenigen Monaten einen Rennwagen mit neuem Motor entwickeln mussten. Der Aufwand zahlte sich aus, denn am 4. Juli 1914 feierte Mercedes im 18/100 Grand-Prix-Rennwagen einen Triumph in Frankreich: Auf dem Rundkurs von Lyon gewann Christian Lautenschlager vor seinen Markenkollegen Louis Wagner und Otto Salzer.

Specifications

Manufacturer:	Daimler
Country:	Germany
Model:	Mercedes 18/100 Grand Prix Racing Car
Produced:	1914
Length:	4100 mm
Width:	1700 mm
Height:	1400 mm
Wheelbase:	2845 mm
Empty weight:	1082 kg
Type of drive:	Rear-wheel drive
Max. speed:	180 km/h
Fuel consumption:	n.a.

The history of the Grand Prix races begins in France. In 1906, the first automobile competition designated as "Grand Prix" took place on a round track in France, organized by the French automobile club. The French Grand Prix soon became the climax of the season in international motorsports. In 1908, the German driver Christian Lautenschlager achieved a spectacular win in Dieppe and thus provided Mercedes with a first Grand Prix success. However, it was not before 1914 that the Mercedes team participated again in the French Grand Prix. In fall 1913, the organizers had issued new technical rules, so Paul Daimler and his engineers had to develop a racing car with a new engine in just a few months. The effort proved worthwhile, when Mercedes triumphed in France on July 4th 1914: On the round track of Lyon, Christian Lautenschlager won before his brand colleagues Louis Wagner and Otto Salzer.

Der Motor

Motordaten

Bauart:	Viertakt-Reihenmotor
Zylinderzahl:	4
Hubraum:	4.483 cm³
Bohrung:	93 mm
Hub:	165 mm
Leistung:	78 kW (105 PS)
	bei 3.100/min

Für den Großen Preis von Frankreich 1914 limitierten die Veranstalter das zulässige Fahrzeuggewicht auf 1.100 Kilogramm, zugleich begrenzten sie den erlaubten Hubraum auf 4,5 Liter. Um bei dem prestigeträchtigen Rennen an den Start gehen zu können, musste die Daimler-Motoren-Gesellschaft deshalb einen neuen Motor für den Mercedes Grand-Prix-Rennwagen entwickeln. Die Ingenieure entschieden sich für einen Reihen-Vierzylinder, der über eine oben liegende Nockenwelle und vier Ventile pro Zylinder verfügte. Der 4,5-Liter-Motor leistete 77 kW (105 PS) bei 3.100/min. Damit erreichte das Mercedes-Triebwerk Drehzahlen, die in der damaligen Zeit als sensationell galten.

Dass der Motor mit der Bezeichnung M93654 trotzdem auch längeren Belastungen gewachsen war, bewies der Sieg von Lautenschlager beim 770 Kilometer langen Grand Prix in Lyon. 1915 gewann der US-Rennfahrer Ralph de Palma mit einem der erfolgreichen Mercedes Grand-Prix-Rennwagen die 500 Meilen von Indianapolis.

The engine

Engine specifications

Type:	Four-stroke in-line engine
Number of cylinders:	4
Displacement:	4483 cm³
Bore:	93 mm
Stroke:	165 mm
Power:	78 kW (105 hp)
	at 3100 rpm

The organizers of the French Grand Prix of 1914 limited the acceptable car weight to 1100 kg and the acceptable displacement to 4.5 liters. In order to be able to participate in this prestigious race, the Daimler company had to develop a new engine for the Mercedes Grand Prix Racing Car. The engineers opted for a 4-cylinder in-line engine with overhead camshaft and four valves per cylinder. The 4.5-liter engine provided 77 kW (105 hp) at 3100 rpm. This engine speed was deemed sensational at its time.

Lautenschlager's Grand Prix win on the 770 km track in Lyon proved that the engine with the designation M93654 could also handle sustained strain. In 1915, the US race driver Ralph de Palma won the Indianapolis 500 with the successful Mercedes Grand Prix Racing Car.

Der Motor des Mercedes Grand-Prix-Rennwagens von 1914 hatte pro Zylinder vier Ventile.

The engine of the Mercedes Grand Prix Racing Car of 1914 had four valves per cylinder.

Der Mercedes 18/100 Grand-Prix-Rennwagen von 1914 besaß nur an der Hinterachse Bremsen.

The Mercedes 18/100 Grand Prix racing car of 1914 was equipped with brakes at the rear axle only.

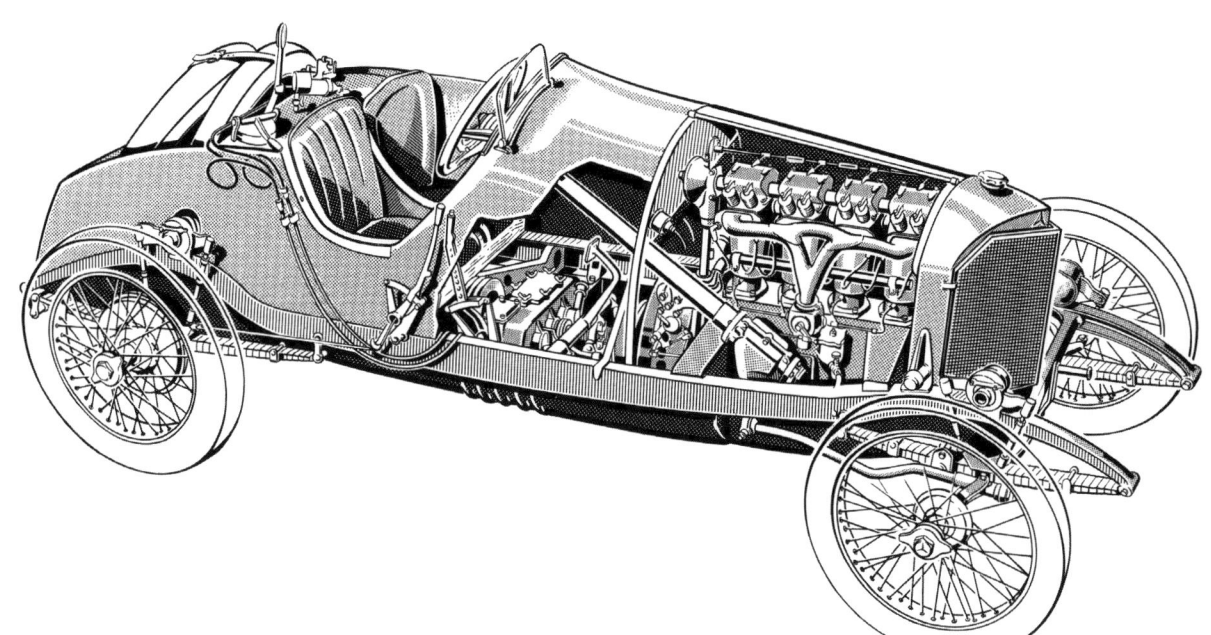

Christian Lautenschlager steuerte den Mercedes Typ 18/100 beim Großen Preis von Frankreich 1914 zum Sieg. Ein Jahr später gewann Ralph de Palma mit einem solchen Modell auch die 500 Meilen von Indianapolis.

Christian Lautenschlager steered the Mercedes Typ 18/100 to victory at the French Grand Prix of 1914. Next year, Ralph de Palma won the Indianapolis 500 with the same model.

Der Peugeot Typ 156 war die letzte Luxuskarosse des französischen Herstellers.
The Peugeot Type 156 was the last luxury car of the French manufacturer.

Peugeot Typ 156

Fahrzeugdaten

Hersteller:	Peugeot
Land:	Frankreich
Modell:	Typ 156
Baujahr:	1921–1923
Länge:	4.800 mm
Breite:	k.A.
Höhe:	k.A.
Radstand:	3.670 mm
Leergewicht:	k.A.
Antriebsart:	Hinterrad
Höchstgeschwindigkeit:	90 km/h
Verbrauch:	k.A.

Der Typ 156 war der erste große von Peugeot nach dem Ersten Weltkrieg produzierte Wagen. Allein sein gewaltiger Radstand von 3,67 Meter verrät, in welcher Liga das Modell 156 angesiedelt war. Von ihm wurden zwischen 1921 und 1923 180 Stück als Limousine, Coupe, Cabrio und Tornado hergestellt. Er war der letzte in Serie hergestellte Peugeot im Luxus-Segment. Ein Peugeot Typ 156 wurde vom französischen Staat für den Fuhrpark des Präsidenten erworben.

Im Peugeot 156 arbeitete in der Serienausstattung ein mächtiger Sechszylinder-Motor mit 6 Litern Hubraum. Mit ihm erreichte der Wagen die für seine Zeit beachtliche Höchstgeschwindigkeit von 90 km/h.

Specifications

Manufacturer:	Peugeot
Country:	France
Model:	Type 156
Produced:	1921–1923
Length:	4800 mm
Width:	n.a.
Height:	n.a.
Wheelbase:	3670 mm
Empty weight:	n.a.
Type of drive:	Rear-wheel drive
Max. speed:	90 km/h
Fuel consumption:	n.a.

The Type 156 was the first large car that Peugeot produced after World War I. The huge wheelbase of 3.67 meters already shows the class to which the model 156 belonged. From 1921 to 1923, 180 units of this type were built as sedan, coupe, convertible and Tornado. This was the last mass-produced Peugeot in the segment of luxury cars. The French government purchased a Peugeot Type 156 for the fleet of the President.

The series model of the Peugeot 156 was driven by a powerful 6-cylinder engine with a displacement of 6 liters. It allowed the car a maximum speed of 90 km/h, which was quite respectable in its time.

Der Motor

Motordaten

Bauart:	Viertakt-Reihenmotor
Zylinderzahl:	6
Hubraum:	5.954 cm³
Bohrung:	95 mm
Hub:	140 mm
Leistung:	18 kW (25 PS)

Das Serienmodell des Peugeot Typ 156 war mit einem 18 kW (25 PS) starken Sechszylinder-Motor mit 6 Litern Hubraum ausgestattet. Peugeot verzichtete auf Ein- und Auslassventile, ihre Aufgabe wurde von Rohrschiebern übernommen. Die neuartige Technik brachte Peugeot 1925 auf den sizilianischen Bergstraßen auf das Podium des legendären Langstreckenrennens Targa Florio. Dennoch blieb die Technologie im Motorenbau nur eine Randerscheinung.

Wegen seiner aerodynamischen Form bot sich der Typ 156 zur Erprobung der ersten Dieselmotoren an. Mit dem „Schwerölmotor" wurde der Wagen zum Wegbereiter der modernen Diesel-Pkw. Der erste dieser Motoren war als Zweizylinder-Zweitakter mit Zündkerze ausgeführt. Er entwickelte 13 kW (18 PS). Bereits auf seiner ersten Testfahrt von Paris nach Bordeaux und zurück erwies sich der Antrieb unter härtesten Bedingungen als robust und zuverlässig. Mit ihm wurde die respektable Höchstgeschwindigkeit von 70 km/h erreicht.

The engine

Engine specifications

Type:	4-stroke in-line engine
Number of cylinders:	6
Displacement:	5954 cm³
Bore:	95 mm
Stroke:	140 mm
Power:	18 kW (25 hp)

The series model of the Peugeot 156 had a 6-cylinder engine with 18 kW (25 hp) and a displacement of 6 liters. Peugeot dispensed with the usual type of intake and exhaust valves and used sleeve valves instead. With this new technology, Peugeot earned a place on the podium at the legendary long-range race Targa Florio along the Sicilian mountain roads in 1925. However, the new technology was only of marginal importance to engine construction.

Due do its aerodynamic shape, the 156 suggested itself to test the first Diesel engines. Equipped with a "heavy oil engine," the vehicle became the trailblazer for the modern Diesel car. The first of these engines was constructed as a 2-stroke 2-cylinder engine with spark plug and was rated at 13 kW (18 hp). On its first test drive from Paris to Bordeaux and back, the engine already proved robust and reliable even in extremely dire circumstances. It achieved a respectable maximum speed of 70 km/h.

In seiner Standardausführung arbeitete im Peugeot 156 ein Sechszylinder-Reihenmotor mit einer Leistung von 18 kW.

In the standard model of the Peugeot 156, a 6-cylinder in-line engine provided 18 kW.

Der Alfa Romeo P2 war ein klassischer Grand-Prix-Rennwagen der 1920er-Jahre.
The Alfa Romeo P2 was a classic Grand Prix racing car of the 1920s.

Alfa Romeo P2 Gran Premio 1925

Fahrzeugdaten

Hersteller:	Alfa Romeo
Land:	Italien
Modell:	P2 Gran Premio
Bauzeit:	1924–1925
Länge:	3.950 mm
Breite:	1.550 mm
Höhe:	1.180 mm
Radstand:	2.624 mm
Leergewicht:	750 kg
Antriebsart:	Hinterrad
Höchstgeschwindigkeit:	225 km/h
Verbrauch:	k.A.

*Werte für 1925 Alfa Romeo P2

Unter der Leitung von Vittorio Jano entstand bei Alfa Romeo 1924 das Modell P2. Dieser Grand-Prix-Rennwagen besaß einen 2-Liter-Achtzylinder-Motor mit Kompressor und war von Beginn an erfolgreich: Antonio Ascari steuerte den P2 beim ersten Einsatz in Cremona sofort zum Sieg.

1925 schrieb die internationale Automobilsportkommission CSI die erste Grand-Prix-Weltmeisterschaft in der Geschichte des Motorsports aus. Es handelte sich um eine WM für Marken, bei der Alfa Romeo mit dem weiterentwickelten P2 antrat. Ascari gewann in Spa-Francorchamps, verunglückte aber beim Großen Preis von Frankreich tödlich. Gastone Brilli-Peri sicherte Alfa Romeo mit seinem Sieg beim italienischen Grand Prix in Monza schließlich den Weltmeistertitel 1925.

Der P2 blieb viele Jahre konkurrenzfähig und bescherte der Mailänder Marke noch bis 1930 weitere Rennerfolge. Insgesamt produzierte Alfa Romeo sechs Exemplare des legendären Rennwagens. Zwei davon existieren noch heute.

Specifications

Manufacturer:	Alfa Romeo
Country:	Italy
Model:	P2 Gran Premio
Produced:	1924–1925
Length:	3950 mm
Width:	1550 mm
Height:	1180 mm
Wheelbase:	2624 mm
Empty weight:	750 kg
Type of drive:	Rear-wheel drive
Max. speed:	225 km/h
Fuel consumption:	n.a.

* data for the 1925 Alfa Romeo P2

Under the direction of Vittorio Jano the Alfa Romeo P2 was created in 1924. This Grand Prix racing car had a 2-liter 8-cylinder engine with compressor and was successful from the very beginning. During its first assignment, Antonio Ascari rode the P2 in Cremona to victory.

In 1925, the international motorsports committee CSI organized the first Grand Prix world championship in the history of motorsports. It was a brand championship, and Alfa Romeo entered the competition with the advanced P2 model. Ascari won in Spa-Francorchamps, but died in an accident at the French Grand Prix. Gastone Brilli-Peri secured the win at the Italian Grand Prix in Monza for Alfa Romeo, and finally the world championship in 1925.

The P2 remained competitive for many years and bestowed many more racing wins upon the Milan brand until 1930. In total, Alfa Romeo produced six units of this legendary racing car, two of which are still in existence.

Der Motor

Motordaten

Bauart:	Viertakt-Reihenmotor mit Kompressor
Zylinderzahl:	8
Hubraum:	1.987 cm³
Bohrung:	61 mm
Hub:	85 mm
Leistung:	114 kW (155 PS) bei 5.500/min

Für den Grand-Prix-Boliden Alfa Romeo P2 entwarf Vittorio Jano einen Reihen-Achtzylinder mit Kompressor. Der 2,0-Liter-Motor, ausgestattet mit zwei Memini-Vergasern, leistete bei seinem Debüt 1924 zunächst 103 kW (140 PS). Ein Jahr später brachte es eine weiterentwickelte Version des Triebwerks auf 114 kW (155 PS) bei 5.500/min. Die Jano-Konstruktion verfügte über zwei oben liegende Nockenwellen sowie zwei Ventile pro Zylinder. Im Grand-Prix-Sport war der erfolgreiche P2 bald nicht mehr startberechtigt, weil sich das technische Reglement änderte. Die Rennmotoren durften ab 1926 nur noch 1,5 Liter Hubraum besitzen – das Triebwerk des Alfa Romeo war damit zu groß. Trotzdem war die Erfolgsgeschichte des P2 noch lange nicht zu Ende, denn er konnte weiterhin bei jenen Rennen eingesetzt werden, die nicht dem Grand-Prix-Reglement unterlagen. 1930 sorgte Achille Varzi mit seinem Sieg bei der berühmten Targa Florio für den letzten Triumph des Alfa Romeo P2.

The engine

Engine specifications

Type:	Four-stroke in-line engine with compressor
Number of cylinders:	8
Displacement:	1987 cm³
Bore:	61 mm
Stroke:	85 mm
Power:	114 kW (155 hp) at 5500 rpm

Vittorio Jano designed an 8-cylinder in-line engine with compressor for the Grand Prix racer Alfa Romeo P2. At its debut in 1924, the 2.0-liter engine with two Memini carburetors delivered 103 kW (140 hp). A year later, and the advanced version already reached 114 kW (155 hp) at 5500 rpm. The Jano design had two overhead camshafts as well as two valves per cylinder. Because of changing technical rules, the successful P2 was soon no longer allowed to enter the Grand Prix. From 1926 on, racing motors were limited to a displacement of 1.5 liters, and hence the engine of the Alfa Romeo was too large. Nonetheless the success story of the P2 did not end for a long time because it could still participate in races that were not governed by the Grand Prix rules. In 1930, Achille Varzi won the famous Targa Florio and thus provided the final triumph of the Alfa Romeo P2.

Der Achtzylinder-Motor mit seinen 114 kW (155 PS) beschleunigte den Alfa Romeo P2 auf bis zu 225 km/h.

The 8-cylinder engine with 114 kW (155 hp) accelerated the Alfa Romeo P2 up to 225 km/h.

Mit dem Modell P2 gewann Alfa Romeo 1925 den ersten Weltmeistertitel in der Geschichte des Grand-Prix-Sports. Der Rennwagen besaß einen 2-Liter-Achtzylinder-Motor, der mit einem Kompressor ausgerüstet war.

In 1925, Alfa Romeo won the first world championship in the history of the Grand Prix sport with the model P2. The racing car was equipped with a 2-liter 8-cylinder engine with compressor.

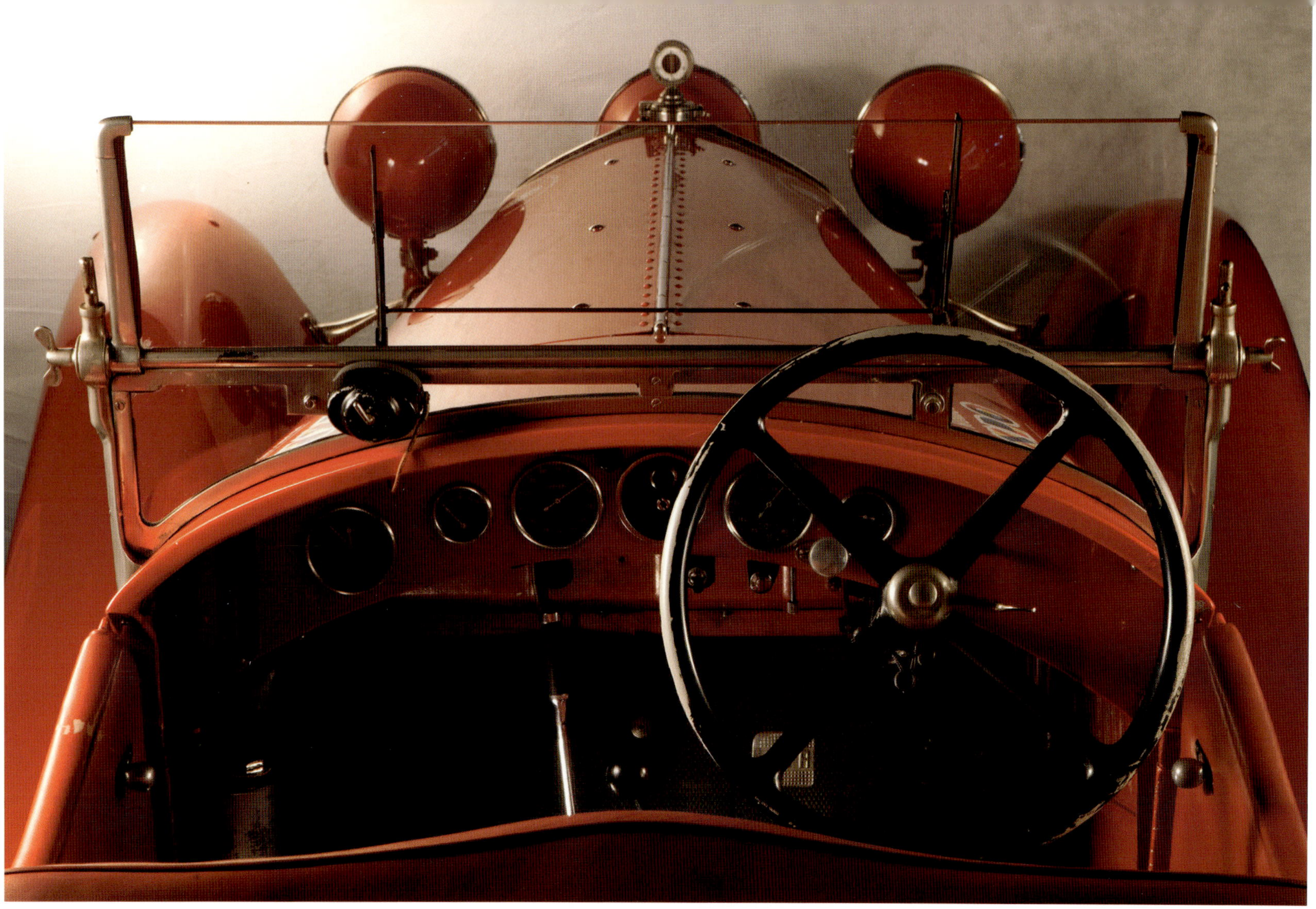

Mit dem Modell 6C 1750 Gran Sport triumphierte Alfa Romeo bei der legendären Mille Miglia.

With the model 6C 1750 Gran Sport, Alfa Romeo triumphed at the legendary Mille Miglia.

Alfa Romeo 6C 1750 Gran Sport 1930

Fahrzeugdaten

Hersteller:	Alfa Romeo
Land:	Italien
Modell:	6C 1750 Gran Sport
Bauzeit:	1929–1930
Länge:	3.652 mm
Breite:	1.615 mm
Höhe:	k.A.
Radstand:	2.745 mm
Leergewicht:	920 kg
Antriebsart:	Hinterrad
Höchstgeschwindigkeit:	145 km/h
Verbrauch:	k.A.

1929 präsentierte Alfa Romeo das Modell 6C 1750 Gran Sport. Wie damals üblich, lieferte das Mailänder Unternehmen seinen Kunden lediglich ein sogenanntes Rolling Chassis – also das Fahrgestell inklusive Motor und Getriebe. Die Fertigung der Karosserie übernahmen Spezialfirmen. Beim 6C 1750 Gran Sport sorgte meist Zagato für die Außenhülle des Spiders.

Der Alfa Romeo besaß einen Reihen-Sechszylinder-Motor, der über einen Hubraum von 1752 cm³ verfügte. Das Triebwerk, das mit einem Kompressor ausgestattet war, leistete 63 kW (85 PS) bei 4.500 Umdrehungen pro Minute. Der 920 Kilogramm schwere Zweisitzer erreichte damit eine Höchstgeschwindigkeit von 145 km/h. Der 6C 1750 Gran Sport feierte im Motorsport zahlreiche Erfolge, darunter auch Gesamtsiege bei der Mille Miglia: 1929 triumphierte Giuseppe Campari im Alfa Romeo, bevor auch Tazio Nuvolari ein Jahr später das Straßenrennen quer durch Italien mit einem 6C 1750 Gran Sport gewann.

Specifications

Manufacturer:	Alfa Romeo
Country:	Italy
Model:	6C 1750 Gran Sport
Produced:	1929–1930
Length:	3652 mm
Width:	1615 mm
Height:	n.a.
Wheelbase:	2745 mm
Empty weight:	920 kg
Type of drive:	Rear-wheel drive
Max. speed:	145 km/h
Fuel consumption:	n.a.

In 1929, Alfa Romeo introduced the 6C 1750 Gran Sport. As it was usual at that time, the Milan company only delivered a so-called rolling chassis, e.g. the undercarriage including engine and transmission, while the auto body was built by specialized companies. With the 6C 1750 Gran Sport, the outer casing was mostly provided by Zagato.

The Alfa Romeo had a 6-cylinder in-line engine with a displacement of 1752 cm³. The engine was equipped with a compressor and delivered 63 kW (85 hp) at 4500 rpm. With this engine, the two-seater of 920 kg reached a top speed of 145 km/h. The 6C 1750 Gran Sport reaped a lot of success in motorsports, e.g. overall wins in the Mille Miglia. In 1929, Giuseppe Campari triumphed driving an Alfa Romeo, and one year later, Tazio Nuvolari also won the Italian cross-country street race in a 6C 1750 Gran Sport.

Der Motor

Motordaten

Bauart:	Viertakt-Reihenmotor mit Kompressor
Zylinderzahl:	6
Hubraum:	1.752 cm³
Bohrung:	65 mm
Hub:	88 mm
Leistung:	63 kW (85 PS) bei 4.500/min

Seit 1927 produzierte Alfa Romeo Fahrzeuge mit der Bezeichnung „6C". Das erste Modell dieser Baureihe hatte einen kleinen Sechszylinder-Motor mit 1,5 Litern Hubraum und brachte Alfa Romeo Sporterfolge. So sicherte sich Giuseppe Campari mit dem 6C 1500 Sport Spider den wichtigen Sieg bei der berühmten Mille Miglia 1928.

Das Nachfolgemodell 6C 1750 Gran Sport hatte ab 1929 einen größeren Hubraum und mehr Leistung. Konstrukteur Vittorio Jano stattete den 1,75 Liter großen Motor zudem mit zwei oben liegenden Nockenwellen aus. Pro Zylinder besaß das wassergekühlte Aggregat ein Ein- und ein Auslassventil. Motorblock und Zylinderkopf des Alfa Romeo waren aus Grauguss gefertigt. Der Wagen verfügte über einen Kompressor, mit dessen Hilfe der Sechszylinder 63 kW (85 PS) bei 4.500 min produzierte. Die Werks-Rennwagen von Alfa Romeo leisteten 75 kW (102 PS). Alle Ausführungen des 6C 1750 Gran Sport übertrugen die Kraft per Viergang-Getriebe auf die Hinterräder.

The engine

Engine specifications

Type:	Four-stroke in-line engine with compressor
Number of cylinders:	6
Displacement:	1752 cm³
Bore:	65 mm
Stroke:	88 mm
Power:	63 kW (85 hp) at 4500 rpm

Since 1927, Alfa Romeo had been producing vehicles with the 6C designation. The first model of this series had a small 6-cylinder 1.5-liter engine and brought Alfa Romeo success in motorsports. With a 6C 1500 Sport Spider, Giuseppe Campari gained the most important win of the famous Mille Miglia 1928.

The successor model 6C 1750 Gran Sport introduced in 1929 had a larger displacement and more power. Constructing engineer Vittorio Jano also had equipped the 1.75-liter engine with two overhead camshafts. The water-cooled engine had one inlet and one outlet valve per cylinder. Engine block and cylinder head of the Alfa Romeo were made of grey cast iron. A compressor helped the 6-cylinder engine to produce 63 kW (85 hp) at 4500 rpm. The Alfa Romeo factory racing cars delivered 75 kW (102 hp). All versions of the 6C 1750 Gran Sport used a four-gear transmission to transfer the power to the rear wheels.

Der Reihen-Sechszylinder des Alfa Romeo leistete 1929 serienmäßig 63 kW (85 PS) bei 4.500 Umdrehungen.

In 1929, the standard 6-cylinder in-line engine of the Alfa Romeo produced 63 kW (85 hp) at 4500 rpm.

Der Alfa Romeo 6C 1750 Gran Sport besaß einen modern gestalteten Sechszylinder-Motor. Konstrukteur Vittorio Jano hatte das Kompressor-Triebwerk 1929 mit zwei oben liegenden Nockenwellen ausgerüstet.

The Alfa Romeo 6C 1750 Gran Sport had a 6-cylinder engine of modern design. Constructing engineer Vittorio Jano had equipped the compressor engine with two overhead camshafts in 1929.

Mercedes-Benz 770 „Großer Mercedes" 1930

Fahrzeugdaten

Hersteller:	Mercedes-Benz
Land:	Deutschland
Modell:	770 „Großer Mercedes"
Bauzeit:	1930–1938
Länge:	5.600 mm
Breite:	1.840 mm
Höhe:	1.830 mm
Radstand:	3.750 mm
Leergewicht:	2.700 kg
Antriebsart:	Hinterrad
Höchstgeschwindigkeit:	160 km/h
Verbrauch:	30 Liter/100 km

Der Mercedes-Benz 770 feierte im Oktober 1930 auf dem Pariser Automobilsalon seine Premiere. Er wurde auch als „Großer Mercedes" bezeichnet. Das repräsentative Topmodell der Stuttgarter Marke verfügte über einen 7,7-Liter-Achtzylinder-Reihenmotor, der wahlweise mit oder ohne Kompressor erhältlich war. Zunächst war der 770 ab Werk nur als Pullmann-Limousine lieferbar. Ab September 1932 ergänzten vier verschiedene Cabriolet-Varianten und ein offener Tourenwagen die Karosserieauswahl. Der Wagen war vorne und hinten mit Starrachsen versehen. Technisch wirkte das Modell mit der internen Bezeichnung „W07" insgesamt eher konservativ. Doch es traf den Geschmack der betuchten Kundschaft, zumal der Wagen auch in gepanzerter Ausführung angeboten wurde. Alle Fahrzeuge wurden ohnehin in Einzelanfertigung individuell auf die Kundenwünsche zugeschnitten. Allein an das japanische Kaiserhaus lieferte Mercedes-Benz bis 1935 sechs Exemplare des Typs 770.

Specifications

Manufacturer:	Mercedes-Benz
Country:	Germany
Model:	770 "Großer Mercedes"
Produced:	1930–1938
Length:	5600 mm
Width:	1840 mm
Height:	1830 mm
Wheelbase:	3750 mm
Empty weight:	2700 kg
Type of drive:	Rear-wheel drive
Max. speed:	160 km/h
Fuel consumption:	30 l/100 km

The Mercedes-Benz 770, also dubbed "Großer Mercedes" ("Large Mercedes"), debuted at the Paris Automobile Salon in October 1930. The prestigious top model of the Stuttgart-based manufacturer had a 7.7-liter 8-cylinder in-line engine, which could be ordered with or without compressor. At first, the 770 was only available as Pullmann sedan ex-factory. Beginning with September 1932, the auto body range was extended by four cabriolet versions and an open touring car. Front and rear axles where fixed. Technically, the model with the internal designation "W07" was rather conservative. However, it suited the taste of the well-to-do customers, in particular as it was also available as an armor-plated version. All vehicles were built individually and tailored to the requirements of the customers. The imperial Japanese household alone purchased six units of the type 770 until 1935.

Der Motor

Motordaten

Bauart:	Viertakt-Reihenmotor mit Kompressor
Zylinderzahl:	8
Hubraum:	7.655 cm³
Bohrung:	95 mm
Hub:	135 mm
Leistung:	146 kW (200 PS) bei 2.800/min

Im „Großen Mercedes" arbeitete ein 7,7-Liter großer Achtzylinder-Reihenmotor. In seiner Grundausführung ohne Kompressor betrug seine Höchstleistung 110 kW (150 PS). Gegen einen Aufpreis von 3.000 Reichsmark war aber auch eine aufgeladene Motor-Variante mit zuschaltbarem Roots-Kompressor lieferbar. Sie leistete 146 kW (200 PS) bei 2.800/min. Von 117 Kunden, die bis Ende 1938 einen Mercedes-Benz 770 bestellten, verzichteten nur 13 auf den Kompressor.

Jeder Zylinder des Motors war mit einem Ein- und einem Auslassventil ausgestattet. Die Ventile wurden über eine seitlich angeordnete Nockenwelle gesteuert. Stirnräder trieben die Nockenwelle an. Der Achtzylinder-Motor mit der internen Bezeichnung „M07" besaß einen Mercedes-Benz-Dreidüsen-Doppelvergaser. Das Verdichtungsverhältnis lag bei 4,7:1. Der Wagen verfügte über eine Zweischeiben-Trockenkupplung. Per Dreigang-Schaltgetriebe mit zusätzlichem Schnellgang wurde die Motorleistung auf die Hinterräder übertragen.

The engine

Engine specifications

Type:	Four-stroke in-line engine with compressor
Number of cylinders:	8
Displacement:	7655 cm³
Bore:	95 mm
Stroke:	135 mm
Power:	146 kW (200 hp) at 2800 rpm

The "Großer Mercedes" was driven by a 7.7-liter 8-cylinder in-line engine. The basic version without compressor had a maximum power of 110 kW (150 hp). For an extra charge of 3000 Reichsmark, a charged motor variant with an on-demand Roots compressor and 146 kW (200 hp) at 2800 rpm was available. Only 13 of the 117 customers who ordered a Mercedes-Benz 770 until the end of 1938 dispensed with the compressor.

Every cylinder of the engine had one intake and one outlet valve. The valves were controlled by a camshaft at the side, which was driven by spur gears. The 8-cylinder engine with the internal designation "M07" had a Mercedes-Benz three-jet twin carburetor. The compression ratio amounted to 4.7:1. The car was equipped with a two-disk dry clutch. The engine power was transferred to the rear wheels via a three-gear transmission with additional fast gear.

Kraftwerk: Der Motor mit der Bezeichnung M07 sorgte im „Großen Mercedes" für Vortrieb.
Power plant: The "Großer Mercedes" was driven by an engine with the designation "M07."

Seltenheit: Dieser Mercedes 770 besaß kein Kompressor-Triebwerk. Der Saugmotor leistete 110 kW (150 PS).
Rarity: This Mercedes 770 did not have a compressor engine. The suction engine provided 110 kW (150 hp).

Der Mercedes-Benz 770 war in den 1930er-Jahren die repräsentative Luxus-Limousine schlechthin für die Reichen und Mächtigen dieser Welt. Allein das japanische Kaiserhaus kaufte sechs Exemplare des „Großen Mercedes".

In the 1930s, the Mercedes-Benz 770 was the prestigious limousine for the rich and the powerful. The Japanese imperial court alone purchased six "Großer Mercedes."

Das Horch 670 Cabriolet beeindruckte 1931 durch elegante Formen und fortschrittliche Technik.
In 1931, the Horch 670 Cabriolet impressed by its elegant shape and advanced technology.

Horch 12 Typ 670 Cabriolet

Fahrzeugdaten

Hersteller:	Horch
Land:	Deutschland
Modell:	12, Typ 670 Cabriolet
Bauzeit:	1931–1934
Länge:	5.400 mm
Breite:	1.820 mm
Höhe:	1.650 mm
Radstand:	3.450 mm
Leergewicht:	2.300 kg
Antriebsart:	Hinterrad
Höchstgeschwindigkeit:	140 km/h
Verbrauch:	ca. 26 Liter/100 km

Die Marke Horch aus dem sächsischen Zwickau war schon in den 1920er-Jahren für ihre Luxusautos berühmt. Im Herbst 1931 präsentierte das Unternehmen auf dem Pariser Automobil Salon eine neue Baureihe mit der Bezeichnung „Horch 12". Herausragendes Merkmal dieser neuen Fahrzeugtypen war ein 6,0-Liter-V12-Motor, der in den Modellen 600 und 670 für Vortrieb sorgte. Zunächst brachte Horch das zweitürige 670 Cabriolet auf den Markt, das über einen Radstand von 3,45 Metern verfügte. Noch imposanter fiel das Modell 600 aus, denn hier wuchs der Radstand auf 375 Zentimeter. Der 600 war ab 1932 als Limousine und Pullman-Cabriolet lieferbar. Die Einführung der Zwölfzylinder-Fahrzeuge aber fiel in die Zeit der Weltwirtschaftskrise. Und obwohl Horch mit bemerkenswerten technischen Lösungen glänzen konnte, blieb der Verkaufserfolg deutlich hinter den Erwartungen zurück. Bis zum Produktionsende 1934 entstanden insgesamt lediglich 81 Exemplare der Baureihe Horch 12.

Specifications

Manufacturer:	Horch
Country:	Germany
Model:	12 Type 670 Cabriolet
Produced:	1931–1934
Length:	5400 mm
Width:	1820 mm
Height:	1650 mm
Wheelbase:	3450 mm
Empty weight:	2300 kg
Type of drive:	Rear-wheel drive
Max. speed:	140 km/h
Fuel consumption:	approx. 26 l/100 km

The Horch brand based in Zwickau, Saxony, was already famous for its luxury cars in the 1920s. In autumn 1931, the company presented the new Horch 12 series at the Paris Motor Show. The distinguishing feature of these new vehicle types was the 6.0-liter V12 engine in the 600 and 670 models. First, Horch launched the two-door convertible 670 with a wheelbase of 3.45 m. With a wheelbase of 375 cm, the 600 model was even more impressive. From 1932 on, the 600 was available as a sedan and a Pullman cabriolet. Unfortunately, the introduction of the 12-cylinder vehicles coincided with the Great Depression. Although Horch could shine with extraordinary technical solutions, the selling figures did not meet the expectations. Until the end of production in 1934, only 81 units of the Horch 12 were built in total.

Der Motor

Motordaten

Bauart:	Viertakt-60-Grad-V-Motor
Zylinderzahl:	12
Hubraum:	6.021 cm³
Bohrung:	80 mm
Hub:	100 mm
Leistung:	88 kW (120 PS)
	bei 3.200/min

Der große Zwölfzylinder-Motor, der ab 1931 im Horch 670 Cabriolet zum Einsatz kam, bestach durch einen ausgesprochen ruhigen Lauf. Konstrukteur Fritz Fiedler hatte den 60-Grad-V12 mit einer siebenfach gelagerten Kurbelwelle ausgestattet, die zusätzlich über zwölf Ausgleichsgewichte und einen Schwingungsdämpfer verfügte. Auch der hydraulische Ventilspielausgleich war ein bemerkenswertes Detail des Horch-Motors. Die Ein- und Auslassventile waren liegend angeordnet und wurden über Kipphebel von der Nockenwelle angetrieben. Der 6,0-Liter-Motor leistete 88 kW (120 PS) bei 3.200/min. Das Viergang-Getriebe des Horch kam aus dem Hause ZF. Und auch hier bot das luxuriöse Fortbewegungsmittel Hightech der 1930er-Jahre: Die Gänge zwei bis vier waren bereits synchronisiert und erleichterten so das Fahren erheblich. Zusätzlich verfügte das Horch 670 Cabriolet über eine Servo-Bremsanlage. Bis heute sind vier Fahrzeuge der Baureihe Horch 12 erhalten.

The engine

Engine specifications

Type:	4-stroke 60° V-engine
Number of cylinders:	12
Displacement:	6021 cm³
Bore:	80 mm
Stroke:	100 mm
Engine power:	88 kW (120 hp)
	at 3200 rpm

The massive 12-cylinder engine used in the Horch 670 Cabriolet of 1931 was distinguished by a markedly smooth run. Its designer, Fritz Fiedler, had equipped the 60° V12 with a crankshaft that was not only mounted on seven points but also had twelve balance weights and an oscillation damper. The hydraulic valve-clearance compensation was another remarkable detail of the Horch engine. The intake and outlet valves were placed horizontally and driven by the camshaft via rocker levers. The 6.0-liter engine provided 88 kW (120 hp) at 3200 rpm. The four-gear transmission of the Horch was built by the ZF company. The luxurious vehicle provided 1930s high-tech also in this area. The second to forth gears were already synchronized, which made driving significantly easier. Additionally, the Horch 670 Cabriolet was also equipped with a servo brake. Today, four Horch 12 vehicles are still in existence.

Der mächtige V12-Motor des Horch hatte 6,0 Liter Hubraum und leistete 88 kW (120 PS) bei 3.200/min.

The powerful V12 Horch engine had a displacement of 6.0 liters and provided 88 kW (120 hp) at 3200 rpm.

Horch präsentierte die luxuriösen Zwölfzylinder-Modelle während der großen Weltwirtschaftskrise. Der Verkaufserfolg hielt sich in Grenzen – nur 81 Exemplare entstanden bis zum Produktionsende 1934.

Horch presented these luxurious 12-cylinder models during the Great Depression. Sales were limited. Until the end of production in 1934, only 81 units had been built.

Markant: Der 303 besaß als erstes BMW-Modell einen Kühlergrill in Form der typischen Doppelniere.
Distinctive trait: The 303 was the first BMW model with a radiator grille in the typical double-kidney shape.

BMW 303

Fahrzeugdaten

Hersteller:	BMW
Land:	Deutschland
Modell:	303
Bauzeit:	1933
Länge:	3.900 mm
Breite:	1.440 mm
Höhe:	1.550 mm
Radstand:	2.400 mm
Leergewicht:	820 kg
Antriebsart:	Hinterrad
Höchstgeschwindigkeit:	90 km/h
Verbrauch:	9–10 Liter/100 km

In der Geschichte der Marke BMW spielte der Typ 303 aus dem Jahr 1933 eine besondere Rolle: Er war der erste Wagen des damals noch jungen Automobilherstellers, den die Ingenieure mit einem Sechszylinder-Reihenmotor ausstatteten; ein Merkmal, das typisch für BMW wurde – genau wie der Kühlergrill in Form einer Doppelniere. Und auch dieses Detail fand sich erstmals beim 303, den BMW im thüringischen Eisenach herstellte.

Der Kleinwagen mit seinem 1,2-Liter Sechszylinder-Motor war ab Werk als Limousine, Cabrio-Limousine und sportliches Cabriolet lieferbar. Die Limousine war die preiswerteste Variante des BMW, sie kostete 3.600 Reichsmark. Auf Basis des BMW 303 produzierten Reutter in Stuttgart und Weinberger in München zudem eigene Zweisitzer.

Der Motor des kleinen BMW leistete 22 kW (30 PS), die über ein Viergang-Getriebe auf die Hinterräder übertragen wurden. Der Wagen war rundum mit Trommelbremsen ausgestattet.

1934 wurde der BMW 303 vom Modell 315 abgelöst.

Specifications

Manufacturer:	BMW
Country:	Germany
Model:	303
Produced:	1933
Length:	3900 mm
Width:	1440 mm
Height:	1550 mm
Wheelbase:	2400 mm
Empty weight:	820 kg
Type of drive:	Rear-wheel drive
Max. speed:	90 km/h
Fuel consumption:	9–10 l/100 km

The type 303 built in 1933 played a special role in the history of the BMW brand. For the first time, the then young automobile manufacturer had fitted a car with a 6-cylinder in-line engine. This should become a typical feature of BMW – just like the double-kidney radiator grille, which also debuted with the 303 built in Eisenach, Thuringia.

The subcompact car with a 1.2-liter 6-cylinder engine was available as sedan, cabriolet sedan and sports cabriolet ex-factory. At 3600 Reichsmark, the sedan was the cheapest version. Reutter in Stuttgart and Weinberger in Munich also produced a two-seater based on the BMW 303.

The engine of the small BMW delivered a power of 22 kW (30 hp), which was transferred to the rear wheels by means of a four-gear transmission. All wheels were fitted with drum brakes.

In 1934, the BMW 303 was replaced by the model 315.

Der Motor

Motordaten

Bauart:	Viertakt-Reihenmotor
Zylinderzahl:	6
Hubraum:	1.173 cm³
Bohrung:	56 mm
Hub:	80 mm
Leistung:	22 kW (30 PS)
	bei 3.500/min

Der erste Sechszylinder-Motor, der bei BMW für ein Automobil entwickelt wurde, hatte 1,2 Liter Hubraum und leistete 22 kW (30 PS) bei 4.000/min. Bevor dieses Antriebsaggregat mit der internen Bezeichnung „M78" aber 1933 in Serie ging, waren bei BMW ab 1931 bereits zwei andere Konstruktionsentwürfe geprüft worden. Doch sie wurden aus unterschiedlichen Gründen verworfen. Schließlich entstand mit dem M78 eine Konstruktion nach dem Baukastenprinzip: Zielsetzung der Ingenieure war die Weiterentwicklung des bereits bewährten Reihen-Vierzylinders zu einem Sechszylinder.

In vielen Konstruktionsmerkmalen entsprach der M78 letztlich dem Vierzylinder-Motor M68: So waren zum Beispiel Hub und Bohrung beider Aggregate identisch. Als wesentliche Neuerung wurden beim neuen Sechszylinder Kurbelgehäuse und Zylinderblock als ein gemeinsames Teil ausgeführt.

Auf Basis des M78 entwickelten die BMW-Techniker später eine Reihe weiterer Motoren mit bis zu zwei Litern Hubraum.

The engine

Engine specifications

Type:	Four-stroke in-line engine
Number of cylinders:	6
Displacement:	1173 cm³
Bore:	56 mm
Stroke:	80 mm
Power:	22 kW (30 hp)
	at 3500 rpm

The first 6-cylinder car engine developed by BMW had a displacement of 1.2 liters and provided 22 kW (30 hp) at 4000 rpm. Series production of the engine with the internal designation "M78" began in 1933. Since 1931, BMW had already tested two other designs, which had been discarded for various reasons. The M78 was built according to a modular design principle. The engineers wanted to develop the previously mentioned 4-cylinder in-line engine into a 6-cylinder engine.

Many design elements of the M78 matched the 4-cylinder engine M68, e.g. stroke and bore. An essential innovation, however, was the design of the crankcase and the cylinder block of the M78 as a single unit.

Based on the M78, BMW later developed several other engines with a displacement of up to 2 liters.

Kurbelgehäuse und Zylinderblock
des Reihen-Sechszylinders bildeten
ein gemeinsames Bauteil.

Crankcase and cylinder block of the
6-cylinder in-line engine formed a
single component.

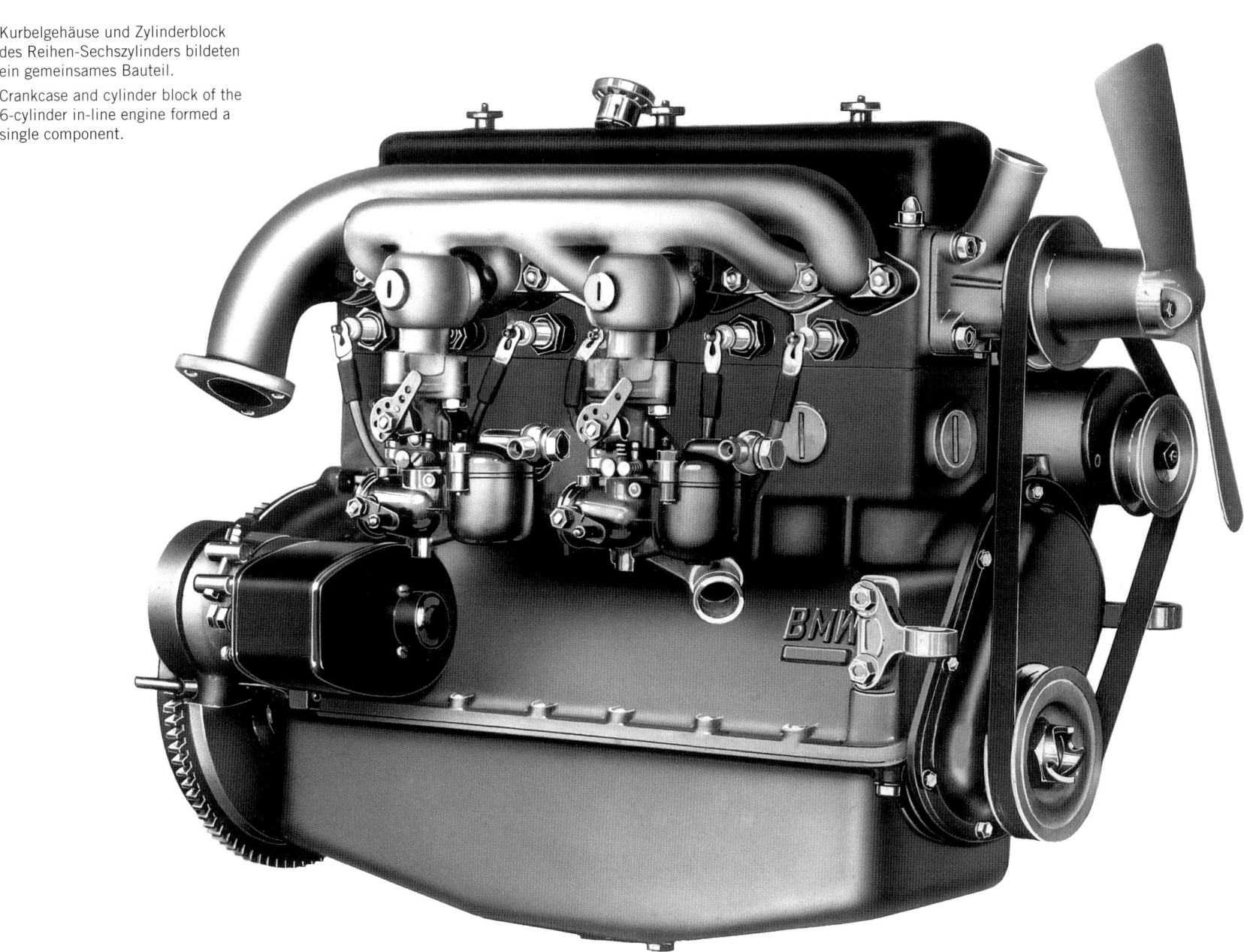

Der erste BMW mit Sechszylinder-Motor entstand 1933 in Eisenach. Im Modell 303 kam ein Triebwerk mit nur 1,2 Litern Hubraum zum Einsatz. 22 kW (30 PS) reichten für eine Höchstgeschwindigkeit von 90 km/h.

The first BMW with a 6-cylinder engine was created in Eisenach in 1933. The model 303 used an engine with a displacement of just 1.2 liters. 22 kW (30 hp) were sufficient to reach a top speed of 90 km/h.

Der Mercedes W25 startete zwischen 1934 und 1936 bei den internationalen Grand-Prix-Rennen.
Between 1934 and 1936, the Mercedes W25 participated in the international Grand Prix races.

Mercedes-Benz W25 „Silberpfeil"

Fahrzeugdaten

Hersteller:	Mercedes-Benz
Land:	Deutschland
Modell:	W25
Bauzeit:	1934–1936
Länge:	4.040 mm
Breite:	1.770 mm
Höhe:	1.160 mm
Radstand:	2.725 mm
Leergewicht:	750 kg ohne Öl, Wasser und Reifen
Antriebsart:	Hinterrad
Höchstgeschwindigkeit:	ca. 280 km/h
Verbrauch:	84–98 Liter/100 km

*Werte für 1934 Mercedes-Benz W25

In der Geschichte des Motorsports gab es zahlreiche Versuche, die immer schnelleren Rennwagen durch Änderungen des technischen Reglements wirkungsvoll einzubremsen. Erfolgreich waren diese Initiativen selten. Doch wohl nie scheiterten die Regelhüter so spektakulär wie bei der Einführung der 750-Kilogramm-Formel für Grand-Prix-Rennen im Jahr 1934. Die PS-Zahlen der Rennwagen explodierten während der Gültigkeit dieses Reglements bis 1937 regelrecht – und die Rundenrekorde fielen auf allen Rennstrecken. Einer der Hauptgründe für den rasanten technischen Fortschritt in jener Zeit war das harte Duell der beiden deutschen Hersteller Mercedes-Benz und Auto Union. Die Schwaben hatten für die neue Rennformel den Typ W25 entwickelt, der beim Eifelrennen 1934 mit Manfred von Brauchitsch am Steuer debütierte. Als erster Rennwagen von Mercedes-Benz präsentierte sich der W25 in einer silbernen Optik – und erhielt dank seiner Erfolge den Spitznamen „Silberpfeil".

Specifications

Manufacturer:	Mercedes-Benz
Country:	Germany
Model:	W25
Produced:	1934–1936
Length:	4040 mm
Width:	1770 mm
Height:	1160 mm
Wheelbase:	2725 mm
Empty weight:	750 kg without oil, water and tires
Type of drive:	Rear-wheel drive
Max. speed:	approx. 280 km/h
Fuel consumption:	84–98 l/100 km

* data for the 1934 Mercedes-Benz W25

In the history of motorsports, there had been numerous attempts to effectively slow down the increasingly fast racing cars by changing the technical rules. However, these initiatives were seldom successful. The rule makers never failed as spectacularly as with the introduction of the 750 kg formula for Grand Prix raced in the year 1934. This rule was in effect until 1937, and during its time, the horsepower figures of the racing cars virtually exploded, and the lap records tumbled on all race tracks. One of the main reasons for the rapid technological advance in this time was the fierce competition between the two German manufacturers Mercedes-Benz and Auto Union. For the new racing formula, the Swabians had developed the type W25, which debuted at the Eifel race with Manfred von Brauchitsch at the wheel. This was the first Mercedes-Benz racing car with a silver appearance. Due to its success, it earned the nickname "Silberpfeil" ("silver arrow").

Der Motor

Motordaten

Bauart:	Viertakt-Reihenmotor mit Kompressor
Zylinderzahl:	8
Hubraum:	3.360 cm³
Bohrung:	78 mm
Hub:	88 mm
Leistung:	260 kW (354 PS) bei 5.800/min

Für den ersten W25 entwickelten die Ingenieure bei Mercedes-Benz ab 1933 einen Achtzylinder-Reihenmotor mit der Bezeichnung „M25A". Er besaß pro Zylinder je zwei schräg hängende Ein- und Auslassventile, die über Schwinghebel betätigt wurden. Zur Ventilsteuerung dienten zwei oben liegende Nockenwellen, die über Stirnräder angetrieben wurden. Die einteilige Kurbelwelle wurde fünffach mit Rollenlagern gelagert. Der Motor war mit einem Roots-Kompressor und zwei Druckvergasern ausgestattet. Das 3,34 Liter große Triebwerk leistete 260 kW (354 PS) bei 5.800/min.

Die Kraft wurde über ein Viergang-Schaltgetriebe auf die Hinterräder übertragen. 1935 gewann Rudolf Caracciola mit dem Mercedes-Benz W25 den EM-Titel.

Die Weiterentwicklung des Motors verhalf dem W25 im Laufe der Jahre zu immer höherer Leistung. Die letzte Variante ME25 verfügte über 4,7 Liter Hubraum und erreichte damit 362 kW (494 PS). Allerdings führten Motorschäden 1936 zu zahlreichen Ausfällen.

The engine

Engine specifications

Type:	Four-stroke in-line engine with compressor
Number of cylinders:	8
Displacement:	3360 cm³
Bore:	78 mm
Stroke:	88 mm
Power:	260 kW (354 hp) at 5800 rpm

Beginning with 1933, the Mercedes-Benz engineers had developed an 8-cylinder in-line engine for the first W25. The engine with the designation "M25A" had two angular overhead inlet and outlet valves per cylinder, which were actuated by swing arms. Two overhead camshafts driven by spur gears were used for valve control. The one-part crankshaft was suspended by roller bearings at five points. The engine was equipped with a Roots compressor and two pressure carburetors. The 3.34-liter engine provided 260 kW (354 hp) at 5800 rpm. The power was transferred to the rear wheels by means of a four-gear transmission. In 1935, Rudolf Caracciola won the European championship aboard a Mercedes-Benz W25.

Further development of the engine increased the power of the W25. The last version, ME25, had a displacement of 4.7 liters and reached 362 kW (494 hp). However, numerous engine failures occurred in 1936.

Der Reihen-Achtzylinder-Motor mit Kompressor leistete 1934 in seiner ersten Version 260 kW (354 PS).

In 1934, the first version of the 8-cylinder in-line engine provided 260 kW (354 hp).

Wartungsarbeiten: Zündkerzenwechsel an zwei Mercedes-Benz W25 vor dem Großen Preis von Deutschland 1934.

Maintenance: Changing the spark plugs of two Mercedes-Benz W25 before the German Grand Prix in 1934.

Weiterentwicklung: Als Typ M25B kam dieser Motor 1935 im Mercedes-Benz W25 zum Einsatz.

Advanced version: In 1935, the M25B engine was used in the Mercedes-Benz W25.

Das Modell W25 begründete ab 1934 den Mythos der Mercedes-Benz-Silberpfeile. Star-Pilot Rudolf Caracciola sicherte sich mit dem Achtzylinder-Kompressor-Rennwagen 1935 den Titel des Grand-Prix-Europameisters.

Beginning with 1934, the W25 model established the legendary reputation of the Mercedes-Benz "silver arrows." With this 8-cylinder compressor racing car, star pilot Rudolf Caracciola secured himself the Grand Prix European championship of 1935.

Der Start zum Großen Preis von Tripolis 1938. Rudolf Caracciola steuerte den Wagen mit der Nummer 26.

Start of the Tripolis Grand Prix in 1938. Rudolf Caracciola drove the car with the number 26.

Rudolf Caracciola nach seinem Sieg beim Großen Preis der Schweiz im August 1938.
Rudolf Caracciola after his win at the Swiss Grand Prix in August 1938.

Caracciola in seinem Mercedes-Benz-Silberpfeil bei einem Boxenstopp.
Caracciola in his Mercedes-Benz silver arrow at a pit stop.

Rudolf Caracciola

Er war der erfolgreichste deutsche Rennfahrer der 1930er Jahre; er begründete mit seinen Erfolgen den legendären Ruf der „Silberpfeile" von Mercedes-Benz: Rudolf Caracciola bewies sein fahrerisches Talent in fast allen Sparten des Motorsports. Er triumphierte bei Grand-Prix-Rennen, siegte im Sportwagen, holte Meistertitel bei Bergrennen, fuhr Geschwindigkeitsweltrekorde und trat sogar bei der Rallye Monte Carlo an. Der berühmte Mercedes-Rennleiter Alfred Neubauer nannte Caracciola „den vielleicht besten Rennfahrer aller Zeiten".

Geboren wurde Rudolf Caracciola am 30. Januar 1901 in Remagen als Sohn einer Hoteliersfamilie. Seine motorsportliche Karriere begann mit Einsätzen auf dem Motorrad, 1922 feierte er den Sieg beim Rennen „Rund um Köln". Bald folgten auch erste Erfolge bei Auto-Rennen. 1923 nahm ihn die Daimler-Motoren-Gesellschaft unter Vertrag – wenn auch zunächst nicht als Rennfahrer. Caracciola arbeitete als Verkäufer in der Daimler Niederlassung im sächsischen Dresden, durfte aber immerhin mit geliehenen Mercedes-Fahrzeugen an Rennen teilnehmen. Dies tat er so erfolgreich, dass er sich schon bald einen Namen als großes Talent machte. Seinen Durchbruch feierte er schließlich 1926 beim ersten Großen Preis von Deutschland auf der Berliner AVUS. Obwohl der Nachwuchspilot seinen Mercedes beim Start abgewürgt und mehr als eine Minute verloren hatte, kämpfte er sich auf regennasser Strecke durch das gesamte Feld und siegte schließlich. Ein Erfolg, der Caracciola den Beinamen „Regenmeister" einbrachte.

Es folgten zahlreiche Erfolge: So siegte er 1927 beim Eröffnungsrennen auf dem neuen Nürburgring und wurde 1930 erstmals Berg-Europameister. 1931 folgte ein historischer Triumph bei der Mille Miglia – als erster Nicht-Italiener siegte der Deutsche bei diesem 1000-Meilen-Rennen.

He was one of the most successful German race drivers of the 1930s, and with his wins, he established the legendary reputation of the Mercedes-Benz "silver arrows": Rudolf Caracciola proved his driving skills in nearly all areas of motorsports. He succeeded in Grand Prix races, in sports cars and in mountain races, he established speed records and even participated in the Monte Carlo Rally. The famous Mercedes race director Alfred Neubauer called Caracciola, "maybe the best race driver of all times."

Rudolf Caracciola was born in a family of hoteliers in Remagen on January 30th 1901. His motorsports career started with motorcycle races. In 1922, he could celebrate a win at the "Around Cologne" race, and soon his first car racing successes followed. In 1923, he was contracted by Daimler; however, not as a race driver. Instead, Caracciola worked as a salesman in the Daimler subsidiary in Dresden, Saxony. He was allowed, though, to participate in races with borrowed Mercedes vehicles. In doing so, he was so successful that he was soon recognized as a great talent. His breakthrough came with the first German Grand Prix on the Berlin AVUS in 1926. Although the budding driver stalled the engine at the start and lost more than a minute, he fought his way through the whole field on a track wet from he rain – and finally won. This success earned him the nickname of "rain master."

Several successes followed: In 1927, he won the opening race on the new Nürburgring, and in 1930, he become European mountain champion for the first time. In 1931, a historic triumph at the 1000-mile race Mille Miglia followed, which he won as the first non-Italian.

Als Mercedes-Benz Ende 1931 alle Rennsportaktivitäten aus wirtschaftlichen Gründen beendete, musste sich Rudolf Caracciola eine Alternative suchen. Er heuerte 1932 bei Alfa Romeo an, gewann erneut die Berg-EM und sicherte sich zudem den Sieg beim Großen Preis von Deutschland. Als sich aber auch Alfa Romeo aus dem Motorsport zurückzog, gründete der Rheinländer 1933 mit dem monegassischen Piloten Louis Chiron ein gemeinsames Team. Beim Training zum Großen Preis von Monaco verunglückte er, weil an seinem Alfa Romeo die Bremsen versagten. Rudolf Caracciola erlitt schwerste Beinverletzungen. Doch schon ein Jahr später kehrte er mit Mercedes-Benz zurück auf die Rennstrecken. Im neuen „Silberpfeil" sicherte er sich 1935 die Grand-Prix-Europameisterschaft – vergleichbar mit der heutigen Formel-1-Weltmeisterschaft. Auch in den Jahren 1937 und 1938 war Caracciola der erfolgreichste Grand-Prix-Pilot und gewann jeweils erneut den Titel. Seinen letzten großen Triumph feierte er 1939 auf dem Nürburgring, als er zum sechsten Mal den Großen Preis von Deutschland gewann.

Der Zweite Weltkrieg unterbrach seine Fahrerkarriere. 1946 sollte er bei den 500 Meilen von Indianapolis in den USA starten. Doch er verunglückte im Training, als sein Kopf vermutlich von einem Vogel getroffen wurde und sein Wagen gegen eine Mauer prallte. Caracciola erlitt Kopfverletzungen und lag einige Tage im Koma. Erst 1952 gab er sein Comeback: Mit dem Mercedes-Benz 300 SL erreichte der inzwischen 51-Jährige bei der Mille Miglia als Vierter das Ziel. Zwei Wochen später trat er beim Großen Preis von Bern an. Mit blockierenden Bremsen kam er von der Strecke ab und kollidierte mit einem Baum. Dabei zog sich Rudolf Caracciola schwere Beinverletzungen zu, die das Ende seiner Karriere bedeuteten. Am 28. September 1959 starb der legendäre Rennfahrer in einer Kasseler Klinik an Leberversagen.

When Mercedes-Benz ceased all motorsports activities for economic reasons at the end of 1931, Rudolf Caracciola had to look for alternatives. In 1932, he signed on at Alfa Romeo, where he won the European mountain championship once again and also the German Grand Prix. When Alfa Romeo also withdrew from motorsports, Caracciola founded his own team together with the Monacan driver Louis Chiron. At the training for the Monaco Grand Prix, he crashed because the brakes of his Alfa Romeo failed. His legs were severely injured. But only one year later, he went back to the race track, this time with a Mercedes-Benz. In the new "silver arrow," he won the Grand Prix European championship of 1935, which is somewhat comparable to the Formula 1 world championship of today. Through the years 1937 and 1938, Caracciola remained the most successful Grand Prix driver and won the championship again.

World War II interrupted his career as a driver. In 1946, he was supposed to participate in the Indianapolis 500; however, he crashed during training when his head was presumably hit by a bird and his car hit a wall. Caracciola's head was injured, and he was in a coma for several days. He had his comeback no sooner than 1952. Now at the age of 51, he made the fourth place at the Mille Miglia, driving a Mercedes-Benz 300 SL. Two weeks later, he participated in the Bern Grand Prix. Due to blocking brakes, he veered off the track and crashed into a tree. He contracted heavy injuries of the legs, which ended his career. The legendary race driver died of liver failure in a Kassel hospital on September 28th 1959.

Rudolf Caracciola – die größten Siege:

1926
Großer Preis von Deutschland

1927
Eifelrennen

1928
Großer Preis von Deutschland
(mit Christian Werner)

1929
Tourist Trophy

1930
Berg-Europameister

1931
Großer Preis von Deutschland
Mille Miglia

Berg-Europameister

1932
Großer Preis von Deutschland
Eifelrennen

Berg-Europameister

1934
Großer Preis von Italien
(mit Luigi Fagioli)

1935
Großer Preis von Belgien
Großer Preis von Spanien
Großer Preis der Schweiz
Großer Preis von Frankreich
Eifelrennen
Großer Preis von Tripolis

Grand-Prix-Europameister

1936
Großer Preis von Monaco

1937
Großer Preis von Deutschland
Großer Preis von Italien
Großer Preis der Schweiz
Masaryk-Grand-Prix

Grand-Prix-Europameister

1938
Großer Preis der Schweiz
Coppa Acerbo

Grand-Prix-Europameister

1939
Großer Preis von Deutschland

Rudolf Caracciola – his biggest wins:

1926
German Grand Prix

1927
Eifel race

1928
German Grand Prix
(with Christian Werner)

1929
Tourist Trophy

1930
European mountain master

1931
German Grand Prix
Mille Miglia

European mountain master

1932
German Grand Prix
Eifel race

European mountain master

1934
Italian Grand Prix
(with Luigi Fagioli)

1935
Belgian Grand Prix
Spanish Grand Prix
Swiss Grand Prix
French Grand Prix
Eifel race
Tripolis Grand Prix

European Grand Prix championship

1936
Monaco Grand Prix

1937
German Grand Prix
Italian Grand Prix
Swiss Grand Prix
Masaryk Grand Prix

European Grand Prix championship

1938
Swiss Grand Prix
Coppa Acerbo

European Grand Prix championship

1939
German Grand Prix

Das beeindruckende Horch 853 Sport-Cabriolet
war ab 1935 auf dem Markt.

The impressive Horch 853 Sport-Cabriolet
was available beginning with 1935.

Bernd Rosemeyer mit seinem Horch 853 Coupé, das eine Karosserie
von Erdmann & Rossi hatte.

Bernd Rosemeyer and his Horch 853 Coupé with an auto body
by Erdmann & Rossi.

Die Konstruktion aus Sachsen stand für luxuriöse Ausstattung und
sportliche Fahrleistungen.

The design from Saxony was an epitome of luxurious equipment and
sporty driving performance.

Horch 853 Sport-Cabriolet 1935

Fahrzeugdaten

Hersteller:	Horch
Land:	Deutschland
Modell:	853 Sport-Cabriolet
Bauzeit:	1935–1939
Länge:	5.300 mm
Breite:	1.830 mm
Höhe:	1.580 mm
Radstand:	3.450 mm
Leergewicht:	2.600 kg
Antriebsart:	Hinterrad
Höchstgeschwindigkeit:	135 km/h
Verbrauch:	22 Liter/100 km

*Werte für 1937 Horch 853 Sport-Cabriolet

Unter den vier Marken der 1932 gegründeten Auto Union ragte Horch hervor: Die Autobauer aus Zwickau besetzten für den neuen Konzern das Luxus-Segment. Und 1935 stellte Horch ein neues Programm mit den zwei Modellreihen 830 und 850 vor. Der Typ 830 besaß einen 3-Liter-V8-Motor, während der Typ 850 mit einem 5-Liter-Reihen-Achtzylinder ausgestattet war.

Topmodell unter den Fahrzeugen der 850er-Baureihe war ein elegantes Sport-Cabriolet mit der Bezeichnung „853", das serienmäßig mit einer De-Dion-Hinterachse ausgerüstet wurde. Anfangs leistete das Triebwerk 73 kW (100 PS), ab 1937 hatte der Achtzylinder 88 kW (120 PS). Damit erreichte der Horch eine Höchstgeschwindigkeit von 135 km/h. Im gleichen Jahr brachten die Sachsen eine überarbeitete Version mit der Bezeichnung 853 A auf den Markt.

Der berühmteste Horch 853 war der Privatwagen des Rennfahrers Bernd Rosemeyer, den die Karosseriefirma Erdmann & Rossi als zweisitziges Coupé gebaut hatte.

Specifications

Manufacturer:	Horch
Country:	Germany
Model:	853 Sport Cabriolet
Produced:	1935–1939
Length:	5300 mm
Width:	1830 mm
Height:	1580 mm
Wheelbase:	3450 mm
Empty weight:	2600 kg
Type of drive:	Rear-wheel drive
Max. speed:	135 km/h
Fuel consumption:	22 l/100 km

* date for the 1937 Horch 853 Sport Cabriolet

Among the four brands of the Auto Union founded in 1932, Horch enjoyed special prominence. The car manufacturer in Zwickau covered the luxury segment of the new corporation. In 1935, Horch presented a new portfolio with the model series 830 and 850. The type 830 had a 3-liter V8 engine, while the 850 was equipped with a 5-liter 8-cylinder in-line engine.

The top model of the 850 series was the elegant sports cabriolet 853, which had a De Dion rear axle in base. In the beginning, the engine delivered 73 kW (100 hp), but from 1937 on, the 8-cylinder engine had 88 kW (120 hp), which provided for a top speed of 135 km/h. In the same year, the Saxon manufacturer launched an advanced version designated as 853A.

The most famous 853 was the private car of the racing driver Bernd Rosemeyer. It had been built as a two-seater coupé by the auto body company Erdmann & Rossi.

Der Motor

Motordaten

Bauart:	Viertakt-Reihenmotor
Zylinderzahl:	8
Hubraum:	4.944 cm³
Bohrung:	87 mm
Hub:	104 mm
Leistung:	90 kW (120 PS)
	bei 3.600/min

In allen Horch-Modellen der Baureihe 850 kam ab Produktionsbeginn ein 5-Liter-Reihen-Achtzylinder-Motor zum Einsatz. Pro Zylinder verfügte das Triebwerk über zwei Ventile, die von einer oben liegenden Nockenwelle gesteuert wurden. Anfangs lag das Verdichtungsverhältnis bei 5,8:1. In dieser Ausführung leistete der Horch-Motor 73 kW (100 PS) bei 3.400 Umdrehungen pro Minute.

Ab 1937 verbaute der sächsische Hersteller eine überarbeitete Version des Antriebsaggregats in den Luxusfahrzeugen. Durch eine modifizierte Nockenwelle und eine höhere Verdichtung stieg die Leistung des wassergekühlten Frontmotors auf 88 kW (120 PS). Das Sport Cabriolet hatte einen Durchschnittsverbrauch von 22 Litern pro 100 Kilometer.

Im Horch 853 war ein Viergang-Getriebe für die Kraftübertragung zuständig. Auf Wunsch war zusätzlich ein sogenannter Schnellgang erhältlich. Dieser Overdrive wurde mit einem zusätzlichen Schalt-Mechanismus aktiviert. 1939 stellte Horch die Produktion des 853 ein.

The engine

Engine specifications

Type:	Four-stroke in-line engine
Number of cylinders:	8
Displacement:	4944 cm³
Bore:	87 mm
Stroke:	104 mm
Power:	90 kW (120 hp)
	at 3600 rpm

Since production begin, all Horch models of the 850 series had a 5-liter 8-cylinder in-line engine with tow valves per cylinder controlled by an overhead camshaft. At first, the compression ratio amounted to 5.8:1. In this version, the Horch engine delivered 73 kW (100 hp) at 3400 rpm.

Beginning in 1937, the Saxon manufacturer used a revised version of the engine in its luxury cars. The modified camshaft and the higher compression rate increased the power of the water-cooled front-mounted engine to 88 kW (120 hp). On average, the Sport Cabriolet consumed 22 liters per 100 km.

In the Horch 853, power was transferred by a four-gear transmission. Optionally, a so-called "fast gear" was also available. This overdrive was activated by an additional switching mechanism. Production of the 853 ended in 1939.

Der 5,0-Liter-Reihenmotor mit seinen acht Zylindern leistete im Horch bis zu 88 kW (120 PS).
The 5.0-liter 8-cylinder in-line engine of the Horch delivered up to 88 kW (120 hp).

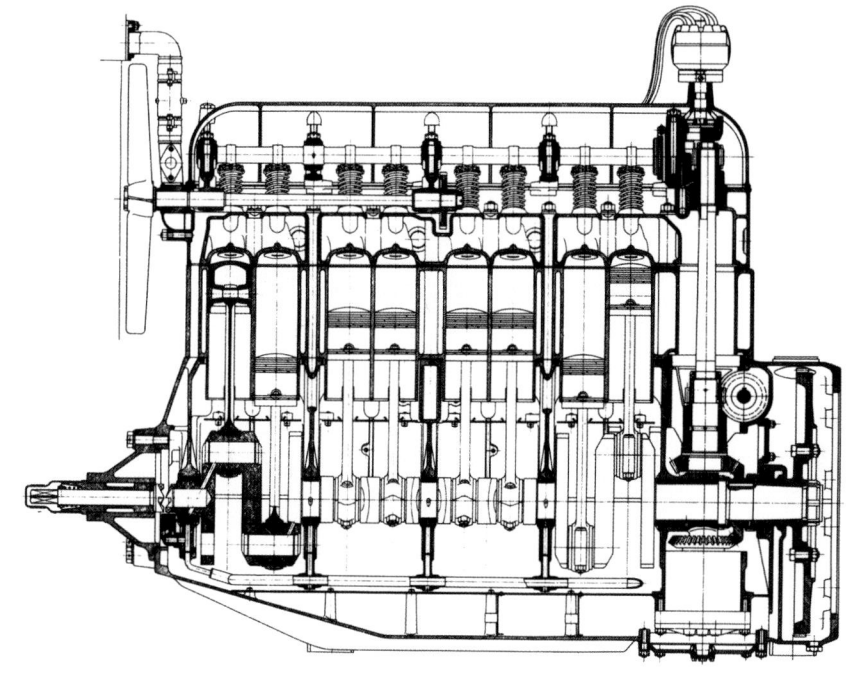

Pro Zylinder besaß der Motor im Horch 853 Sport-Cabriolet zwei Ventile.
The engine of the Horch 853 Sport Cabriolet had two valves per cylinder.

Das Horch 853 Sport-Cabriolet besaß einen imposanten 5,0-Liter-Reihen-Achtzylinder, der bis zu 88 kW (120 PS) leistete. Serienmäßig verfügte das Top-Modell der sächsischen Marke über eine De-Dion-Hinterachse.

The Horch 853 Sport Cabriolet had an impressive 5.0-liter 8-cylinder in-line engine with up to 88 kW (120 hp). The top model of the Saxon brand was equipped with a De Dion rear axle in base.

Der Alfa Romeo 8C-35 war ein
Grand-Prix-Rennwagen, der ab 1935 zum
Einsatz kam.

The Alfa Romeo 8C-35 was a Grand Prix
racing car used from 1935 on.

Alfa Romeo 8C-35

Fahrzeugdaten

Hersteller:	Alfa Romeo
Land:	Deutschland
Modell:	8C-35
Bauzeit:	1935
Länge:	4.200 mm
Breite:	1.520 mm
Höhe:	1.215 mm
Radstand:	2.750 mm
Leergewicht:	750 kg
Antriebsart:	Hinterrad
Höchstgeschwindigkeit:	ca. 275 km/h
Verbrauch:	k.A.

Ab 1934 traten die neuen Rennwagen von Mercedes-Benz und Auto Union bei den Grand-Prix-Rennen in Europa an. Die deutschen Konstruktionen erwiesen sich sehr schnell als eine schwer zu schlagende Konkurrenz für die zuvor dominierenden Marken aus Italien und Frankreich. Um gegen die „Silberpfeile" eine Chance zu haben, entwickelte Alfa Romeo 1935 den 8C-35. Der Wagen besaß einen 3,8-Liter-Achtzylinder-Motor mit Kompressor. Sein Renndebüt feierte die Neukonstruktion beim Großen Preis von Italien: René Dreyfus und Tazio Nuvolari steuerten den Monoposto in Monza auf Platz zwei. Schon 1936 löste eine 12-Zylinder-Version den 8C-35 ab. Trotzdem feierte Tazio Nuvolari im Vorjahreswagen noch zwei Erfolge über die deutsche Konkurrenz: Sowohl in Budapest wie auch in Livorno lag er mit dem Alfa Romeo 8C-35 vorne.

Bei einer Auktion erreichte das Siegerauto von Nuvolari 2013 einen Preis von über sieben Millionen Euro. Damit war er der weltweit teuerste Alfa Romeo.

Specifications

Manufacturer:	Alfa Romeo
Country:	Italy
Model:	8C 35
Produced:	1935
Length:	4200 mm
Width:	1520 mm
Height:	1215 mm
Wheelbase:	2750 mm
Empty weight:	750 kg
Type of drive:	Rear-wheel drive
Max. speed:	approx. 275 km/h
Fuel consumption:	n.a.

Beginning with 1934, the new racing cars of Mercedes-Benz and Auto Union participated in the Grand Prix races in Europe. The German models soon became undefeatable competitors for the previously dominating brands from Italy and France. In order to stand up against the "silver arrow," Alfa Romeo developed the 8C-35 in 1935. This car had a 3.8-liter 8-cylinder engine with compressor and debuted at the Italian Grand Prix, where René Dreyfus and Tazio Nuvolari achieved second place with the open-wheeled racer. In 1936, the 8C-35 was already replaced by a 12-cylinder version. Nonetheless Tazio Nuvolari and the prior model still managed to triumph twice over the German competitors. In Budapest as well as in Livorno Nuvolari succeeded with his Alfa Romeo 8C-35.

At an auction in 2013, his winning car was sold for more than 7 million Euros and thus became the world's most expensive Alfa Romeo.

Der Motor

Motordaten

Bauart:	Viertakt-Reihenmotor mit Kompressor
Zylinderzahl:	8
Hubraum:	3.822 cm³
Bohrung:	78 mm
Hub:	100 mm
Leistung:	246 kW (330 PS) bei 5.400/min

Schon bei seinem ersten Renneinsatz im Herbst 1935 war der Alfa Romeo 8C-35 erkennbar untermotorisiert. Ein neuer Zwölfzylinder-Motor wurde bei Alfa Romeo zwar bereits entwickelt, doch noch mussten sich die Piloten mit dem 246 kW (330 PS) starken Reihen-Achtzylinder mit Kompressor begnügen. Das Triebwerk, das von Vittorio Jano konstruiert worden war, besaß zwei Ventile pro Zylinder, zwei oben liegende Nockenwellen und zwei Weber-Vergaser. Motorblock und Zylinderkopf waren aus Aluminium gefertigt. Die Leistung wurde mit einem Viergang-Getriebe auf die Hinterräder übertragen. Gegen die überlegenen PS-Zahlen der „Silberpfeile" war der Alfa auf schnellen Strecken chancenlos. Doch auf engen Kursen konnte der Achtzylinder durchaus mithalten – wenn Tazio Nuvolari am Steuer saß. Die Erfolge des 8C-35 gingen so vor allem auf das Konto des Ausnahmetalents aus Mantua. Die Scuderia Ferrari war damals übrigens für den Einsatz der Rennwagen von Alfa Romeo verantwortlich.

The engine

Engine specifications

Type:	Four-stroke in-line engine with compressor
Number of cylinders:	8
Displacement:	3822 cm³
Bore:	78 mm
Stroke:	100 mm
Power:	246 kW (330 hp) at 5400 rpm

The Alfa Romeo 8C-35 already proved to be underpowered at its first race in autumn 1935. Alfa Romeo was already developing a new 12-cylinder engine but for the time being, the drivers had to be content with the 8-cylinder in-line engine with 246 kW (330 hp). This engine had been designed by Vittorio Jano and had two valves per cylinder, two overhead camshafts and two Weber carburetors. Engine block and cylinder head were made of aluminum. The power was transferred to the rear wheels via a four-gear transmission. On fast tracks, the Alfa did not stand a chance against the superior engine power of the "silver arrows," but on tight tracks, the 8-cylinder engine could keep up – at least when Tazio Nuvolari was at the wheel. The success of the 8C-35 was mainly due to the exceptional talent from Mantua. Incidentally, at that time, the Scuderia Ferrari was responsible for the deployment of Alfa Romeo racing cars.

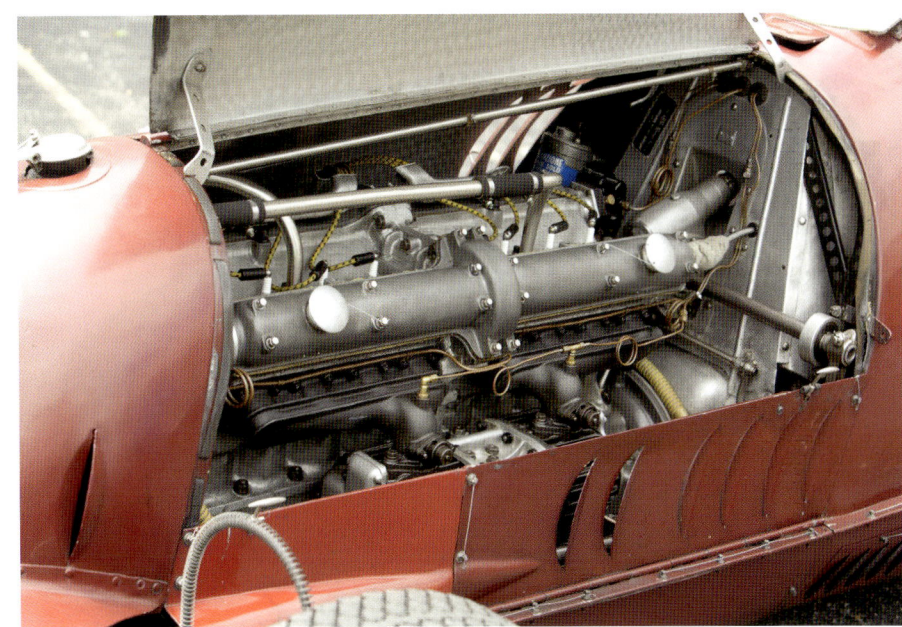

Der 3,8-Liter-Achtzylinder-Reihenmotor mit Kompressor war mit nur 246 kW (330 PS) der Konkurrenz unterlegen.

The 3.8-liter 8-cylinder in-line engine with compressor provided only 246 kW (330 hp) and was thus inferior to the competitors.

Der Alfa Romeo 8C-35 war ein Rennwagen aus dem Jahr 1935. Bei einer Auktion erreichte ein solcher Monoposto 2013 einen Preis von über sieben Millionen Euro. Damit war er der teuerste Alfa Romeo weltweit.

The Alfa Romeo 8C-35 was a racing car built in 1935. At an auction in 2013, one of these open-wheeled racers was sold for more than 7 million Euros and thus became the world's most expensive Alfa Romeo.

Den mit einem Reihensechszylinder bestückten Wanderer W25K gab es als Cabriolet und als Roadster.
The Wanderer W25K with a 6-cylinder in-line engine was available as a cabriolet and as a roadster.

Wanderer W25K

Fahrzeugdaten

Hersteller:	Wanderer
Land:	Deutschland
Modell:	W25K
Bauzeit:	1936–1938
Länge:	4.220 mm
Breite:	1.680 mm
Höhe:	1.400–1550 mm
Radstand:	2.650 mm
Leergewicht:	1.000–1.070 kg
Antriebsart:	Hinterrad
Höchstgeschwindigkeit:	145 km/h
Verbrauch:	20 Liter/100 km

Im Frühjahr 1936 wurde das Modell zunächst mit Kompressor und 85 PS präsentiert. Nun verfügte auch Wanderer über einen offenen Wagen, der das sportliche Image der Auto Union zu festigen half. Der sowohl als Sport-Cabriolet wie auch als Roadster erhältliche W25K wurde zu einem Aushängeschild der Auto Union. Auf Betreiben der Sächsischen Staatsbank, die ihr Engagement im sächsischen Automobilbau in Gefahr sah, schlossen sich bereits am 29. Juni 1932 die Audi Werke, die Horch Werke und die Zschopauer Motorenwerke/DKW zur Auto Union AG zusammen. Gleichzeitig wurde mit den Wanderer Werken im benachbarten Chemnitz ein Kauf- und Pachtvertrag zur Übernahme der Wanderer-Automobilabteilung abgeschlossen. Sitz dieses neuen Konzerns wurde Chemnitz – und als Symbol dieses Zusammenschlusses standen vier Ringe, die bis heute das Audi-Markenzeichen verkörpern.

Auf Kundenwunsch war der wunderschön designte W25K auch mit einem Sportmotor ohne Kompressoraufladung erhältlich.

Specifications

Manufacturer:	Wanderer
Country:	Germany
Model:	W25K
Produced:	1936–1938
Length:	4220 mm
Width:	1680 mm
Height:	1400–1550 mm
Wheelbase:	2650 mm
Empty weight:	1000–1070 kg
Type of drive:	Rear-wheel drive
Max. speed:	145 km/h
Fuel consumption:	20 l/100 km

In spring 1936, this model was first presented with a compressor and a 85-hp engine. Now Wanderer also had an open car model, which helped to the sporty image of Auto Union. The W25K was available both as a sports cabriolet and as a roadster and became one of the flagships of Auto Union. Instigated by the Saxon State Bank, which saw its participations in the Saxon automobile manufacturing at risk, Audi, Horch and Zschopauer Motorenwerke/DKW merged to form the Auto Union AG in June 29th, 1932. At the same time, a lease and selling contract with the Wanderer factories in the neighboring city of Chemnitz was signed to take over the automotive department of Wanderer. The headquarters of the new group was located in Chemnitz. The four rings that make up the brand logo of Audi up to this date were a symbol of this merger.

Owing to customer requests, the beautifully designed W25K was also available with a sports engine without compressor charging.

Der Motor

Motordaten

Bauart:	Viertakt-Reihenmotor mit Kompressor
Zylinderzahl:	6
Hubraum:	1.950 cm³
Bohrung:	70 mm
Hub:	85 mm
Leistung:	62,5 kW (85 PS) bei 5.000/min

Der Sechszylinder-Reihenmotor des Wanderer W25K entstammte ursprünglich dem Wanderer W17 beziehungsweise W22 und war in seinen Grundzügen bereits 1932 von Ferdinand Porsche im Auftrag von Wanderer entwickelt worden. In der zivilen Serienausführung leistete der mit OHV-Steuerung ausgestattete Zwei-Liter-Motor 29 kW (40 PS). Mit Kompressoraufladung, Spezialzylinderkopf und Sportauspuff erreichte er jedoch mehr als das Doppelte und sorgte für dynamische Fahrleistungen. Ergänzt um diese technischen Zutaten, stieg die Leistung auf beachtliche 62,5 kW (85 PS). Auf Kundenwunsch war der Wanderer W25K auch mit einem Sportmotor ohne Kompressoraufladung erhältlich. In dieser Version sorgten eine erhöhte Verdichtung und bearbeitete Ein- und Auslasskanäle für lebendige 55 PS und 120 km/h Höchstgeschwindigkeit. Mangelnde Zuverlässigkeit des Kompressortriebwerks veranlasste Wanderer, das Modell 1938 nur noch mit dem Standardmotor anzubieten.

The engine

Engine specifications

Type:	4-stroke in-line engine with compressor
Number of cylinders:	6
Displacement:	1950 cm³
Bore:	70 mm
Stroke:	85 mm
Engine power:	62.5 kW (85 hp) at 5000 rpm

The 6-cylinder in-line engine of the Wanderer W25K had already been used in the Wanderer W17 and W22. Its basic design was developed by Ferdinand Porsche on behalf of Wanderer in 1932. In the civilian serial version, the 2-liter engine with OHV control was rated at 29 kW (40 hp). With compressor charging, special cylinder head and sports exhaust, it achieved more than twice the power and provided for a dynamic driving performance. These technical additions increased the power to a whopping 62.5 kW (85 hp). Owing to customer requests, the Wanderer W25K was also available with a sports engine without compressor charging. In this version, increased compression and adapted intake and outlet ports provided for a vivid engine power of 55 hp and a top speed of 120 km/h. However, the unreliability of the compressor drive prompted Wanderer to offer the car model only with a standard engine in 1938.

Das geschwungene Instrumentenbrett des W25K war wie der ganze Wagen höchst liebevoll gestaltet.

Like the whole car, the curved dashboard of the W25K was designed with much care.

Der Sechszylinder-Motor des W25K wurde Anfang der 1930er-Jahre von Ferdinand Porsche entwickelt.

The 6-cylinder engine of the W25K was developed by Ferdinand Porsche in the beginning of the 1930s.

Die Grundkonstruktion des Sechszylinder-Reihenmotors im Wanderer W25K ging auf die Arbeiten von Ferdinand Porsche zurück. Mit Kompressor brachte es der 2-Liter-Motor auf bis zu 62,5 kW (85 PS).

The basic construction of the 6-cylinder in-line engine of the Wanderer W25K harked back to a design by Ferdinand Porsche. With the addition of a compressor, the 2-liter engine provided up to 62.5 kW (85 hp).

Diesel-Premiere: Der Mercedes-Benz 260 D von 1936 in seiner Ausführung als Pullman-Limousine.
Diesel debut: The Pullman sedan version of the 1936 Mercedes-Benz 260 D.

Mercedes-Benz Diesel Typ 260 D

Fahrzeugdaten

Hersteller:	Mercedes-Benz
Land:	Deutschland
Modell:	260 D
Bauzeit:	1936–1940
Länge:	4.390 mm
Breite:	1.630 mm
Höhe:	1.600 mm
Radstand:	3.050 mm
Leergewicht:	k.A.
Antriebsart:	Hinterrad
Höchstgeschwindigkeit:	90 km/h
Verbrauch:	k.A.

Der im Februar 1936 auf der Automobilausstellung in Berlin vorgestellte Mercedes-Benz 260 D entsprach äußerlich dem 200er-Mercedes jener Tage. Unter der Motorhaube aber fand sich eine Sensation: ein Dieselmotor. Er machte den 260 D zum ersten in Serie gebauten Diesel-Personenwagen weltweit. Für das Auto sprach der günstige Betrieb: Im Vergleich zum gleichwertigen Typ 230 mit Benzinmotor ließen sich die Treibstoffkosten mit dem Dieselantrieb um 60 Prozent reduzieren. Der 260 D, der auch als W138 bekannt wurde, war mit einem Viergang-Getriebe und hydraulisch betätigten Trommelbremsen an allen vier Rädern ausgestattet. Das Fahrzeug wurde in unterschiedlichen Karosserie-Varianten angeboten: als Limousine, als Cabriolet und als Landaulet. Der Diesel-Mercedes kam häufig als Taxi zum Einsatz. Hier konnte er seine hervorragende Langlebigkeit eindrucksvoll unter Beweis stellen. Bis zum Produktionsende 1940 wurden vom 260 D 1.967 Stück gebaut.

Specifications

Manufacturer:	Mercedes-Benz
Country:	Germany
Model:	260 D
Produced:	1936–1940
Length:	4390 mm
Width:	1630 mm
Height:	1600 mm
Wheelbase:	3050 mm
Empty weight:	n.a.
Type of drive:	Rear-wheel drive
Max. speed:	90 km/h
Fuel consumption:	n.a.

Introduced at the Berlin Automobile Fair in February 1936, the Mercedes-Benz 260 D looked like a usual Mercedes 200. Under the hood, however, a sensation lurked: a Diesel engine, which made the 260 D the world's first mass produced passenger car. The cheap operation made this model attractive: in comparison to the equivalent Type 230 with gasoline engine, the Diesel engine reduced the fuel costs by 60 %. The 260 D, also known as W138, came with a 4-gear transmission and hydraulically operated drum brakes. The car was offered with various types of auto bodies, as sedan, as convertible and as landaulet. The Diesel Mercedes was often used as a taxi. In this area, it could prove its excellent longevity. Until end of production in 1940, 1967 units of the 260 D were built.

Der Motor

Motordaten

Bauart:	Viertakt-Diesel-Reihenmotor
Zylinderzahl:	4
Hubraum:	2.545 cm³
Bohrung:	90 mm
Hub:	100 mm
Leistung:	33 kW (45 PS)
	bei 3.000/min

Bei Lastkraftwagen hatte der Dieselantrieb seine Bewährungsprobe bereits bestanden, als Mercedes-Benz erste Versuche mit Dieselmotoren bei Personenwagen durchführte. Die Schwaben setzten 1933 dafür einen Reihen-Vierzylinder-Dieselmotor ein. Er hatte einen Hubraum von 3,8 Litern und brachte es bei 2.800 Umdrehungen pro Minute auf 59 kW (80 PS). Dieser Motor neigte bei höheren Drehzahlen aber zu enormen Schwingungen, die in den Testwagen zu Rahmenbrüchen führten. Ohnehin waren Dieselmotoren deutlich schwerer als Benziner und boten beim Beschleunigen nur eine geringe Elastizität. Erst nachdem diese Nachteile auf ein erträgliches Maß verringert werden konnten, startete ab 1936 die Serienfertigung von Pkw-Dieselmotoren. Der OM 138 genannte Vierzylinder-Reihenmotor des Mercedes-Benz 260 D leistete 33 kW (45 PS) bei 3.000 Umdrehungen pro Minute. Sein Verdichtungsverhältnis lag bei 20,5:1. Das „OM" der Motor-Typenbezeichnung stand übrigens für „Öl-Motor".

The engine

Engine specifications

Type:	4-stroke Diesel in-line engine
Number of cylinders:	4
Displacement:	2545 cm³
Bore:	90 mm
Stroke:	100 mm
Power:	33 kW (45 hp)
	at 3000/min

The Diesel engine for trucks had already proven its worth when Mercedes-Benz conducted first tests with Diesel engines for passenger cars. In 1933, the Swabian manufacturers used a 4-cylinder in-line Diesel engine with a displacement of 3.8 liters and 59 kW (80 hp) at 2800 rpm. However, with higher engine speeds, this engine tended to undergo heavy vibrations, which caused frame fractures in the test cars. The Diesel engines were also much heavier than the gasoline engine and did not provide for much flexibility when accelerating. After the problems had been reduced to acceptable dimensions, the series production of passenger car Diesel engines started in 1936. The OM 138 4-cylinder in-line engine of the Mercedes-Benz 260 D provided 33 kW (45 hp) at 3000 rpm. Its compression rate amounted to 20.5:1. The letters "OM" in the type designation stood for "oil motor."

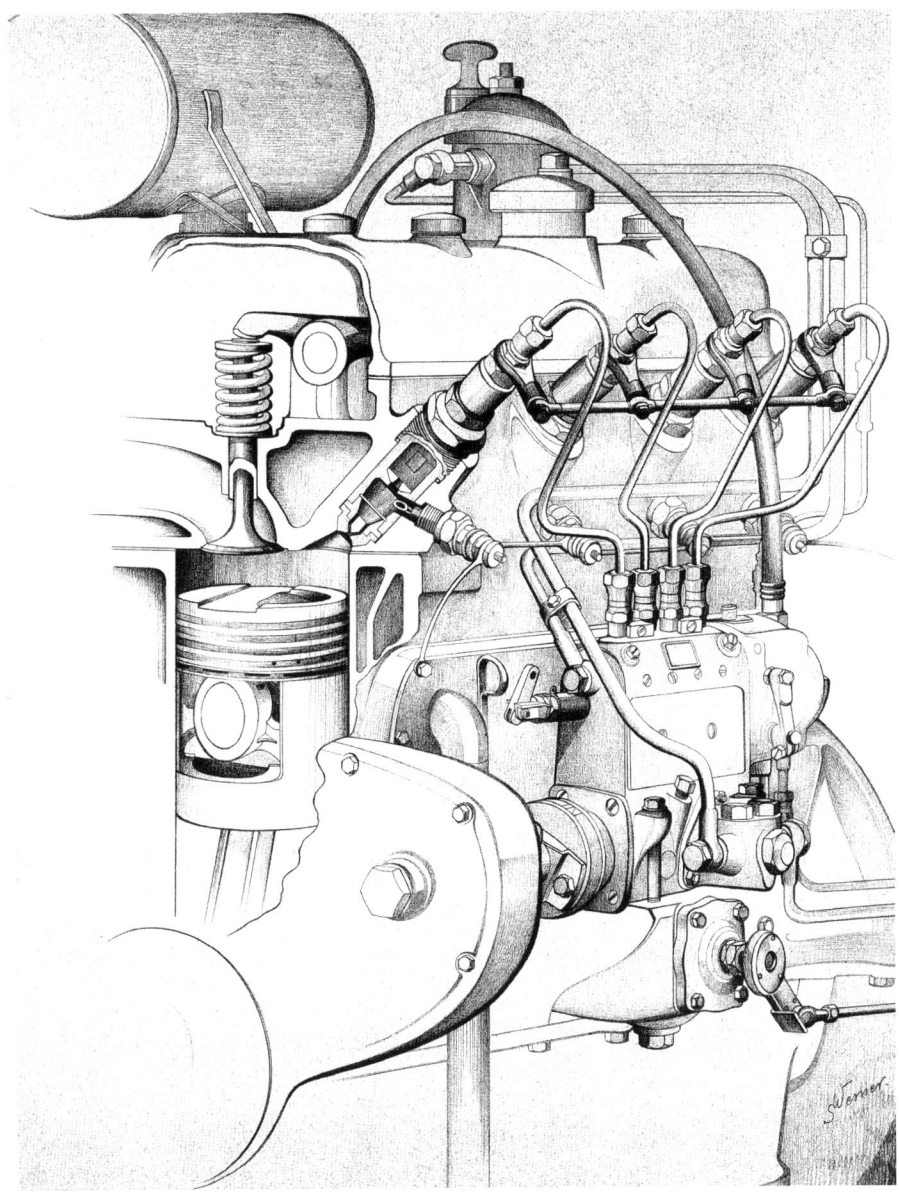

1936 leistete der Vierzylinder-Dieselmotor OM 138 im Mercedes-Benz 260 D 33 kW (45 PS) bei 3.000/min.

In 1936, the 4-cylinder Diesel engine OM 138 in the Mercedes-Benz 260 D provided 33 kW (45 hp) at 3000/min.

Als erster Personenwagen der Welt besaß der Mercedes-Benz 260 D serienmäßig einen Dieselmotor.

The Mercedes-Benz 260 D was the world's first passenger car with a Diesel engine in base.

Mit dem 260 D begann 1936 eine neue Epoche, denn der Mercedes-Benz war der erste in Serie gebaute Personenwagen mit Dieselmotor. Dieser Antrieb überzeugte durch niedrige Unterhaltskosten und Langlebigkeit.

In 1936, the 260 D started a new era as this Mercedes-Benz was the first mass-produced passenger car with a Diesel engine. Low cost of ownership and longevity made this engine very compelling.

Beeindruckende Kulisse: der Mercedes-Benz W125 beim Großglockner Grand Prix 2012.
Impressive backdrop: Mercedes-Benz W125 at the Grossglockner Grand Prix 2012.

Mercedes Benz W125 „Silberpfeil"   1937

Fahrzeugdaten

Hersteller:	Mercedes-Benz
Land:	Deutschland
Modell:	W125
Bauzeit:	1937
Länge:	4.200 mm
Breite:	1.750 mm
Höhe:	1.200 mm
Radstand:	2.798 mm
Leergewicht:	749 kg ohne Öl, Wasser und Reifen
Antriebsart:	Hinterrad
Höchstgeschwindigkeit:	ca. 320 km/h
Verbrauch:	ca. 100 Liter/100 km

Nachdem Mercedes-Benz in der Saison 1936 im Kampf gegen das Team der Auto Union fast chancenlos geblieben war, musste ein neuer Rennwagen her. Die Arbeit an der Neukonstruktion begann im August 1936. Der Nachfolger des Mercedes-Benz W25 trug den Namen „W125" – und setzte neue Maßstäbe. Mit 434 kW (592 PS) war der Monoposto der stärkste Grand-Prix-Rennwagen, den es bis dahin gegeben hatte. Diesen Rekord hielt der W125 über viele Jahrzehnte. Erst die Turbo-Formel-1-Rennwagen der 1980er Jahre konnten schließlich mit mehr Leistung protzen als dieser „Silberpfeil".

Der W125 war 1937 der Konkurrenz überlegen. Der Grund dafür war nicht allein die hohe Motorleistung, die deutlich über den Werten der Konkurrenz lag. Im Vergleich zum Vorgänger W25 verfügte die Neukonstruktion zudem über einen längeren Radstand, neue Radaufhängungen und ein deutlich verwindungssteiferes Chassis. Rudolf Caracciola gewann im Mercedes-Benz W125 souverän den Europameister-Titel.

Specifications

Manufacturer:	Mercedes-Benz
Country:	Germany
Model:	W125
Produced:	1937
Length:	4200 mm
Width:	1750 mm
Height:	1200 mm
Wheelbase:	2798 mm
Empty weight:	749 kg without oil, water and tires
Type of drive:	Rear-wheel drive
Max. speed:	approx. 320 km/h
Fuel consumption:	approx. 100 l/100 km

When Mercedes-Benz did not stand a chance against the Auto Union team in the 1936 season, a new racing car had to be developed. Work on the new design commenced in August 1936. The successor of the Mercedes-Benz W25 was named "W125" and set a new benchmark. At 434 kW (592 hp), this open-wheel car was the most powerful Grand Prix racing car until that time and for several decades after. Only the Turbo Formula 1 racing cars of the 1980s could boast more power than this "silver arrow."

In 1937, the W125 was superior to its competitors. This was not only due to the high engine power, which significantly exceeded the specifications of the competitors. The new design also had a wheelbase that was much longer than that of its predecessor W25 and a much more warp resistant chassis. With the Mercedes-Benz W125, Rudolf Caracciola won the European championship with aplomb.

Der Motor

Motordaten

Bauart:	Viertakt-Reihenmotor mit Kompressor
Zylinderzahl:	8
Hubraum:	5.660 cm³
Bohrung:	94 mm
Hub:	102 mm
Leistung:	434 kW (592 PS) bei 5.800/min

Im Mercedes Benz W125 kam ein großer Reihen-Achtzylinder mit Roots-Kompressor zum Einsatz, der grundsätzlich der Konstruktion aus dem Vorgängermodell W25 ähnelte. So bestand auch dieser Motor aus zwei Stahlgussblöcken mit jeweils vier Zylindern und einem aufgeschweißten Stahlblechmantel für das Kühlwasser. Das Achtzylinder-Aggregat mit der Bezeichnung „M125" war 222 Kilogramm schwer, hatte rund 5,7 Liter Hubraum und brachte es bei 5.800 Umdrehungen pro Minute auf beeindruckende 434 kW (592 PS). Die Höchstdrehzahl lag bei 6.000/min. Jeder Zylinder des M125 verfügte über je zwei schräg hängende Ein- und Auslassventile, die über Schwinghebel betätigt wurden. Die Ventilsteuerung erfolgte über zwei oben liegende Nockenwellen. Das Verdichtungsverhältnis betrug 8,8:1. Im Laufe der Saison 1937 wurde der Motor des Mercedes-Benz W125 stetig weiterentwickelt – in seiner letzten Evolutionsstufe erreichte er bei Prüfstandmessungen eine Leistung von 474 kW (646 PS).

The engine

Engine specifications

Type:	Four-stroke in-line engine with compressor
Number of cylinders:	8
Displacement:	5660 cm³
Bore:	94 mm
Stroke:	102 mm
Power:	434 kW (592 hp) at 5800 rpm

The Mercedes Benz W125 was driven by a large 8-cylinder in-line engine with Roots compressor similar to the design used in the predecessor model W25. This engine also was made of two cast steel blocks with eight cylinders each and a welded-on steel sheet jacket for the cooling water. The engine with the designation "M125" weighed 222 kg, had a displacement of 5.7 liters and delivered an impressive 434 kW (592 hp) at 5800 rpm. The maximum engine speed amounted to 6000 rpm. Each of the cylinders of the M125 had two angular overhead intake and outlet valves actuated by swing arms. Two overhead camshafts were used for valve control. The compression ratio was 8.8:1. During the 1937 season, the engine of the Mercedes-Benz W125 was continually improved. In the last step of its evolution, it showed a power of 474 kW (646 hp) in test measures.

Kraftprotz:
Der Mercedes-Benz W125
war der stärkste Renn-
wagen der Silberpfeil-Ära.

Muscle man:
The Mercedes-Benz W125
was the most powerful
racing car of the
"silver arrow" era.

Unverhüllt: Chassis und
Motor des Mercedes-Benz
W125 ohne Karosserie.

Unconcealed:
Chassis and engine of the
Mercedes-Benz W125
without auto body.

Der 5,6-Liter-
Achtzylinder war der
stärkste Rennmotor der
Vorkriegszeit.

The 5.6-liter 8-cylinder
engine was the most
powerful pre-war racing
engine.

Der Mercedes-Benz W125 aus dem Jahr 1937 besaß einen 5,6-Liter-Achtzy-linder-Motor. Das Kompressor-Triebwerk leistete 434 kW (592 PS), was den Boliden zum stärksten Grand-Prix-Wagen seiner Epoche machte.

The 1937 Mercedes-Benz W125 had a 5.6-liter 8-cylinder engine with compressor, which provided 434 kW (592 hp). This made the W125 the most powerful Grand Prix car of its era.

Bernd Rosemeyer beim Start zu einer Rekordfahrt auf der Autobahn bei Frankfurt 1937.

Bernd Rosemeyer at the start for a record ride on the autobahn near Frankfurt in 1937.

Auto Union Typ C Rekordwagen 1937

Fahrzeugdaten

Hersteller:	Auto Union
Land:	Deutschland
Modell:	Typ C Weltrekordwagen
Bauzeit:	1937
Länge:	5.800 mm
Breite:	1.950 mm
Höhe:	1.110 mm
Radstand:	2.910 mm
Leergewicht:	910 kg
Antriebsart:	Hinterrad
Höchstgeschwindigkeit:	410 km/h
Verbrauch:	k.A.

Rekordfahrten gehörten seit 1934 zum festen Programm der Auto-Union-Rennabteilung. Auf Basis ihrer Grand-Prix-Rennwagen bauten die Sachsen immer schnellere Autos, die diverse Bestmarken aufstellten. Mit dem Typ-C-Stromlinien-Rennwagen entstand 1937 ein Modell, das zunächst bei einem Renneinsatz im Mai 1937 für Furore sorgte: Im Training zum AVUS-Rennen in Berlin drehte Bernd Rosemeyer im Auto Union eine Runde mit einer Durchschnittsgeschwindigkeit von 284,3 km/h.

Mit Motorproblemen verpasste der Star der Auto Union im Endlauf als Vierter nur knapp das Siegerpodium. Mehr Grund zum Feiern hatte der Deutsche wenige Tage später, als er mit dem Stromlinien-Renner auf der Autobahn Frankfurt–Darmstadt einen Geschwindigkeitsweltrekord aufstellte. Im Oktober 1937 fuhr Bernd Rosemeyer schließlich an gleicher Stelle mit dem Rekordauto 406,3 km/h schnell und stellte damit einen neuen Rekord auf.

Specifications

Manufacturer:	Auto Union
Country:	Germany
Model:	Type C world record car
Produced:	1937
Length:	5800 mm
Width:	1950 mm
Height:	1110 mm
Wheelbase:	2910 mm
Empty weight:	910 kg
Type of drive:	Rear-wheel drive
Max. speed:	410 km/h
Fuel consumption:	n.a.

Since 1934, record rides had become a regular activity of the Auto Union racing division. Based on its Grand Prix racing car, the Saxon manufacturer built increasingly faster cars, which set several records. The streamlined type C racing car caused a sensation at its first outing in May 1937 when Bernd Rosemeyer completed a lap with an average speed of 284.3 km/h during the training for the Berlin AVUS race.

Due to engine problems, the star of the Auto Union only achieved the 4th place in the final run and thus only narrowly missed the podium. A few days later, however, the German driver could celebrate a new speed world record with the streamlined racing car on the Frankfurt–Darmstadt autobahn. In October 1937, Bernd Rosemeyer drove the record car at the same spot as fast as 406.3 km/h and thus set a new record.

Der Motor

Motordaten

Bauart:	Viertakt-45-Grad-V-Motor mit Kompressor
Zylinderzahl:	16
Hubraum:	6.005 cm³
Bohrung:	75 mm
Hub:	85 mm
Leistung:	382 kW (520 PS) bei 5000/min

Das Antriebsaggregat des Rekordwagens mit Stromlinienkarosserie war identisch mit dem Motor des Auto-Union-Typ-C-Grand-Prix-Boliden. In beiden Modellen kam ein 6,0-Liter-V16-Zylinder zum Einsatz, der 382 kW (520 PS) bei 5.000 Umdrehungen pro Minute leistete. Das V-Triebwerk hatte einen Zylinderwinkel von 45 Grad. Eine konstruktive Besonderheit des Rennmotors war die Ventilsteuerung durch eine per Königswelle angetriebene zentrale Nockenwelle. Pro Zylinder verfügte der Motor über zwei Ventile. Für die Gemischbildung sorgten zwei Solex-Vergaser, zudem besaß der Auto-Union-Motor einen Roots-Kompressor. Das maximale Drehmoment des Triebwerks betrug beeindruckende 853 Nm bei 2.500/min. Der V16 war als Mittelmotor im Fahrzeug angeordnet.

Für die Rekordversuche im Januar 1938 überarbeitete die Auto Union das Stromlinien-Modell. Durch eine größere Bohrung verfügte der Motor in dieser letzten Version über 6,5 Liter Hubraum und eine Leistung von 410 kW (560 PS).

The engine

Engine specifications

Type:	Four-stroke 45° V engine with compressor
Number of cylinders:	16
Displacement:	6005 cm³
Bore:	75 mm
Stroke:	85 mm
Power:	382 kW (520 hp) at 5000 rpm

The engine of the record car with streamlined auto body was identical to the one of the Auto Union type C Grand Prix car. Both models used a 6.0-liter V16 engine with 382 kW (520 hp) at 5000 rpm. The V engine had a banking angle of 45°. As a special feature of the racing motor, the valves were controlled by a central camshaft driven by an upright shaft. The engine had two valves per cylinder. The mixture was provided by two Solex carburetors. Furthermore, the Auto Union engine was equipped with a Roots compressor. The maximum torque amounted to an impressive 853 Nm at 2500 rpm. The V16 was mounted as a midship engine.

For record-breaking attempts in January 1938, Auto Union modified the streamline model. Larger bores in the latest version of the engine provided for a displacement of 6.5 liters and a power of 410 kW (560 hp).

Der 6,0-Liter-V16-Zylinder aus dem Grand-Prix-Boliden wurde auch im Rekordwagen benutzt.
The 6.0-liter V16 engine of the Grand Prix racing car was also used in the record car.

Ein Blick unter die Karosserie des Rekordwagens vor dem AVUS-Rennen 1937.
A view under the auto body of the record car before the AVUS race in 1937.

Auf Basis des Grand-Prix-Rennwagens Typ C baute die Auto Union diesen Weltrekordwagen. Bernd Rosemeyer erreichte mit dem 16-Zylinder-Boliden auf der Autobahn Frankfurt-Darmstadt im Herbst 1937 über 400 km/h.

Auto Union built this world record car based on the Grand Prix racing car type C. With this 16-cylinder racer, Bernd Rosemeyer reached a speed of more than 400 km/h on the Frankfurt–Darmstadt autobahn in autumn 1937.

Bernd Rosemeyer war einer der größten Grand-Prix-Stars der Vorkriegszeit. 1936 gewann er den Europameisterschafts-Titel.

Bernd Rosemeyer was one of the greatest Grand Prix stars of the pre-war era. In 1936, he won the European championship.

Rosemeyer fuhr auch Weltrekorde: 1937 erreichte er als erster Fahrer mehr als 400 km/h auf einer normalen Straße.

Rosemeyer also set world records: In 1937, he was the first driver to reach more than 400 km/h on a regular road.

Bernd Rosemeyer

Die Geschichte des Bernd Rosemeyer fasziniert Motorsportfans bis heute: Er erlebte als Rennfahrer einen kometenhaften Aufstieg zum Grand-Prix-Star, war das Idol einer ganzen Nation – und starb mit nur 28 Jahren bei einem Unfall.

Geboren wurde er am 14. Oktober 1909 in Lingen. In der kleinen Stadt im Emsland betrieb sein Vater eine Werkstatt für Automobile und Motorräder – so kam der junge Rosemeyer schon früh in Kontakt mit dem Thema Technik. Wie die meisten Rennfahrer seiner Generation begann auch er seine Motorsport-karriere auf zwei Rädern. Ab 1930 ging er bei Motorrad-Rennen an den Start, zeigte schnell überragendes Talent und wurde schon 1932 von NSU als Werks-fahrer verpflichtet. Ein Jahr später wechselte er zu DKW. Diese Marke gehörte zur Auto Union; dem Nachwuchsfahrer boten sich dort also Chancen für den erhofften Wechsel in den Automobilsport.

Die Auto Union beteiligte sich seit 1934 mit einem von Ferdinand Porsche konstruierten Monoposto an Grand-Prix-Rennen. Im Oktober 1934 durfte Bernd Rosemeyer tatsächlich an Testfahrten mit dem Grand-Prix-Boliden auf dem Nürburgring teilnehmen. Auto-Union-Rennleiter Willy Walb war von den Leistungen des völlig unerfahrenen Nachwuchsfahrers offensichtlich beein-druckt und nahm Rosemeyer schließlich unter Vertrag.

Sein Renndebüt im Auto Union feierte Bernd Rosemeyer im Mai 1935 auf der Berliner AVUS, aber schon nach drei Runden musste er seinen Wagen mit Reifenschaden abstellen. Doch nur wenige Wochen später sorgte der 25-Jäh-rige bei seinem zweiten Einsatz für eine Sensation: Beim Eifelrennen auf dem Nürburgring übernahm er drei Runden vor Schluss die Führung von Merce-des-Star Rudolf Caracciola. Erst in der letzten Runde konnte sich Routinier Caracciola doch noch durchsetzen und knapp vor Rosemeyer gewinnen.

Motorsports fans are captivated by the story of Bernd Rosemeyer until today. In a meteoric career, the race driver became a Grand Prix star and the idol of a whole nation. He died in an accident, aged only 28.

He was born in Lingen on October 14th 1909. His father ran a small workshop for cars and motorcycles in the small town, and thus Bernd became exposed to technology at an early age. Similar to most of the race drivers of his gen-eration, his career in motorsports began on two wheels. From 1930 on, he participated in motorcycle races where he soon showed outstanding talent. As early as 1932, NSU contracted him as a factory driver. A year later, he joined DKW, a brand belonging to Auto Union, which offered the budding driver the chance of entering car races.

Since 1934, Auto Union took part in Grand Prix races with an open-wheeled car designed by Ferdinand Porsche. Bernd Rosemeyer participated in test drives of this Grand Prix racer on the Nürburgring in October 1934. Willy Walb, racing director of Auto Union, was impressed by the performance of the completely unexperienced new driver and contracted him.

Bernd Rosemeyer debuted as Auto Union race driver on the Berlin AVUS in May 1935, but he had to stop his car after no more than three laps because of damaged tires. However, only a few weeks later the 25 year old driver caused a sensation at his second outing: during the Eifel race on the Nürburgring, he gained the lead from Mercedes star Rudolf Caracciola three laps before the finish. Not before the last gap, the experienced driver Caracciola could gain the lead again and score a tight win.

Mit seinem bemerkenswerten Auftritt in der Eifel gab der Auto-Union-Pilot eine erste Kostprobe seiner virtuosen fahrerischen Fähigkeiten. Schon im September 1935 feierte er beim Masaryk-Grand-Prix im tschechischen Brünn seinen ersten Sieg. Bei der Siegerehrung dort lernte er die bekannte Fliegerin Elly Beinhorn kennen. Sie war in Deutschland ein Superstar – und wenige Monate später Rosemeyers Ehefrau. Elly und Bernd wurden zum populären Glamour-Paar des „Dritten Reichs". Dazu passten die sportlichen Erfolge des jungen Ehemanns: 1936 dominierte er den Grand-Prix-Sport, feierte Siege in Serie und wurde Europameister.

1937 schlug Mercedes-Benz zurück: Die Schwaben hatten mit dem neuen Modell W125 einen überlegenen Rennwagen auf die Räder gestellt, dem die Auto Union nur wenig entgegensetzen konnte. Allein Bernd Rosemeyer schaffte es, die Dominanz der Star-Mannschaft von Mercedes-Benz zumindest gelegentlich zu brechen. Er siegte beim Eifelrennen, gewann den Vanderbilt Cup in New York sowie die Coppa Acerbo in Italien und triumphierte beim Saisonfinale im britischen Donington Park. Es sollte zugleich sein letztes Rennen sein.

Das Duell zwischen Mercedes-Benz und Auto Union wurde nicht nur auf den Rennstrecken ausgetragen, denn ebenso wichtig wie Rennsiege waren Rekordfahrten. Nach Ende der Rennsaison fand im Oktober 1937 auf der Autobahn Frankfurt/Main-Darmstadt die offizielle Rekordwoche der Obersten Nationalen Sportbehörde (ONS) statt. Die Auto Union feierte dort einen prestigeträchtigen Erfolg: In einem Typ-C-Rennwagen mit Vollverkleidung legte Bernd Rosemeyer den fliegenden Kilometer mit 406,3 km/h zurück und stellte damit einen neuen Weltrekord auf. Noch nie zuvor war ein Pilot auf einer normalen Straße schneller als 400 km/h gefahren.

Diese Niederlage wollten die Verantwortlichen bei Mercedes-Benz nicht auf sich sitzen lassen. Obwohl die ONS Rekordfahrten eigentlich nur noch im Herbst hatte zulassen wollen, sollte das Duell schon zu Beginn des Jahres 1938 in seine nächste Runde gehen. Beide Hersteller brachten weiter verbesserte Fahrzeuge zu den Rekordversuchen am 28. Januar 1938: Rudolf Caracciola legte am frühen Morgen im Mercedes-Benz vor und stellte mit 432,7 km/h einen neuen Weltrekord über den fliegenden Kilometer auf. Bernd Rosemeyer startete um 11.46 Uhr zu seiner Rekordfahrt, die fatal endete. Der Auto Union überschlug sich auf der Autobahn Frankfurt-Darmstadt aus ungeklärter Ursache bei Tempo 430. Rosemeyer wurde aus dem Wagen geschleudert und war sofort tot.

With his remarkable performance in the Eifel, the Auto Union driver had shown a taste of his virtuoso driving skills. In September 1935, he could already celebrate his first win at the Masaryk Grand Prix in the Czech city of Brno. At the award ceremony, he met the famous aviator Elly Beinhorn who was a super star in Germany and few months later became Rosemeyer's wife. Elly and Bernd became a popular glamour couple of the Third Reich. The sports success of the young husband was only apt: in 1936, he dominated the Grand Prix sports, gained a string of wins and became European champion.

In 1937, Mercedes-Benz stroke back: with the new model W125, the Swabian manufacturer had put a superior racing car on the tracks. Auto Union had not much to match this car. Only Bernd Rosemeyer managed to overcome the dominance of the Mercedes-Benz star teams at least occasionally. He won the Eifel race, the Vanderbilt Cup in New York and the Coppa Acerbo in Italy and triumphed at the season finale in Donington Park, UK. However, this turned out to be his last race.

The duel between Mercedes-Benz and Auto Union was not only fought on race tracks, because record rides were as important as racing wins. After the end of the racing season, the official 'records week' of the supreme national sports office (Oberste Nationale Sportbehörde, ONS) took place on the Frankfurt/Main–Darmstadt autobahn in October 1937. Here, Auto Union gained a prestigious success: in a type C racing car with full fairing, Bernd Rosemeyer completed the kilometer-lancé at 406.3 km/h and thus set a new world record. Never before had someone driven faster than 400 km/h on a regular street.

However, Mercedes-Benz would not take this defeat sitting down. Although the ONS wanted to allow record rides only in autumn, the duel should already continue at the beginning of 1938. Both manufacturers used improved vehicles attempt to set a new record on January 28th 1938. Rudolf Caracciola started early in the morning in a Mercedes-Benz and set a new world record of 432,7 km/h for the kilometer-lancé. Bernd Rosemeyer started at 11.46 a.m., but his record attempt ended fatally. For unknown reasons, his Auto Union car overturned on the Frankfurt–Darmstadt autobahn at a speed of 430 km/h. Rosemeyer was hurled out of the car and died instantly.

Bernd Rosemeyer – die größten Siege:

1935
Großer Preis der Tschechoslowakei
(Masaryk-Grand-Prix)

1936
Großer Preis von Deutschland
Großer Preis von Italien
Großer Preis der Schweiz
Coppa Acerbo
Eifelrennen

Großer Bergpreis von Deutschland

Grand-Prix-Europameister

1937
Großer Preis von Donington
Coppa Acerbo
Eifelrennen
Vanderbilt Cup

Bernd Rosemeyer – his biggest wins:

1935
Masaryk Grand Prix
(Czechia)

1936
German Grand Prix
Italian Grand Prix
Swiss Grand Prix
Coppa Acerbo
Eifel race

German Mountain Grand Prix

European Grand Prix Champion

1937
Grand Prix of Donington
Coppa Acerbo
Eifel race
Vanderbilt Cup

1937 gewann Bernd
Rosemeyer mit seinem
Auto Union Typ C die Coppa
Acerbo in Pescara.

In 1937, Bernd Rosemeyer
won the Coppa Acerbo
in Pescara in his
Auto Union type C.

Die Karosserie der Cabrio-let-Variante des Audi 920 lieferte die Firma Gläser aus Dresden.

The auto body of the cabriolet version of the Audi 920 was manufactured by the Gläser company in Dresden.

Uhrensammlung: Das Armaturenbrett des Audi 920 verfügte über mittig angeord-nete Anzeigen.

Clock collection: the dash-board of the Audi 920 had dials arranged in the center.

Audi 920

Fahrzeugdaten

Hersteller:	Audi
Land:	Deutschland
Modell:	920
Bauzeit:	1938–1940
Länge:	4.895 mm
Breite:	1.720 mm
Höhe:	1.610 mm
Radstand:	3.100 mm
Leergewicht:	1.640 kg
Antriebsart:	Hinterrad
Höchstgeschwindigkeit:	130 km/h
Verbrauch:	14–16 Liter/100 km

Seit 1932 gehörte Audi gemeinsam mit den Marken Horch, Wanderer und DKW zur neu gegründeten Auto Union. Innerhalb des Konzerns war Audi für Fahrzeuge der gehobenen Mittelklasse zuständig – und damit zwischen der Luxus-Marke Horch und Wanderer platziert.

Ende 1938 präsentierte Audi den Typ 920, der im Zentralen Entwicklungs- und Konstruktionsbüro der Auto Union entstanden war. Die letzte Neukonstruktion, die Audi vor Beginn des Zweiten Weltkriegs auf den Markt brachte, war ein Produkt des Baukastensystems der Auto Union. Karosserie und Fahrgestell wurden vom Wanderer W23 übernommen. Damit verbunden war die Rückkehr zum Heckantrieb, während die Vorgängermodelle des 920 über Frontantrieb verfügt hatten. Den Sechszylinder-Motor mit oben liegender Nockenwelle hatten die Motorentechniker der Auto Union ursprünglich für einen geplanten Horch entwickelt. Bis zum Produktionsende 1940 entstanden insgesamt nur 1.281 Fahrzeuge des Modells 920.

Specifications

Manufacturer:	Audi
Country:	Germany
Model:	920
Produced:	1938–1940
Length:	4895 mm
Width:	1720 mm
Height:	1610 mm
Wheelbase:	3100 mm
Empty weight:	1640 kg
Type of drive:	Rear-wheel drive
Max. speed:	130 km/h
Fuel consumption:	14–16 l/100 km

Since 1932, Audi belonged to the newly founded Auto Union, together with Horch, Wanderer and DKW. In this group, Audi was responsible for cars of the upper medium-class and thus placed between the luxury brand Horch and Wanderer.

In the end of 1938, Audi presented the type 920, which had been developed in the central development and construction office of the Auto Union. This was the last new design that Audi brought to market before the outbreak of World War II. It was a product of the modular construction system of the Auto Union. Auto body and undercarriage were taken from the Wanderer W23. This also meant reverting to a rear-wheel drive, while the predecessor models of the 920 had used a front-wheel drive. The 6-cylinder engine with overhead camshaft had originally been developed for the Horch. Until end of production in 1940, only 1281 units the model 920 were built.

Audi 920

Der Motor

Motordaten

Bauart:	Viertakt-Reihenmotor
Zylinderzahl:	6
Hubraum:	3.281 cm³
Bohrung:	87 mm
Hub:	92 mm
Leistung:	55 kW (75 PS)
	bei 3.000/min

Der zwischen 1938 und 1940 produzierte Audi 920 besaß einen Reihen-Sechszylinder-Motor. Grundsätzlich basierte er auf der Achtzylinder-Konstruktion des Horch 850. Das wassergekühlte Aggregat verfügte über 3,3 Liter Hubraum und leistete 55 kW (75 PS) bei 3.000 Umdrehungen pro Minute. Der Motor war mit zwei Ventilen pro Zylinder und einer oben liegenden Nockenwelle mit Königswellenantrieb ausgerüstet. Die Konstrukteure der Auto Union hatten das Triebwerk zudem mit einem Solex-Registriervergaser bestückt. Der 920 besaß ein Viergang-Getriebe, das die Kraft des Motors auf die Hinterräder übertrug.

Der Wagen erreichte eine Höchstgeschwindigkeit von 130 km/h, doch dieser Wert galt nur als kurzzeitige Spitzengeschwindigkeit. Auf den neuen Reichsautobahnen zeigte sich, dass viele Motoren der damaligen Zeit noch nicht für längere Vollgasfahrten geeignet waren. Und so gab auch die Auto Union 1938 für den Audi die „Autobahn-Dauergeschwindigkeit" mit nur 118 km/h an.

The engine

Engine specifications

Type:	Four-stroke in-line engine
Number of cylinders:	6
Displacement:	3281 cm³
Bore:	87 mm
Stroke:	92 mm
Power:	55 kW (75 hp)
	at 3000 rpm

The Audi 920 was built between 1938 and 1940 had a 6-cylinder in-line engine based on the 8-cylinder design of the Horch 850. The water-cooled engine had a displacement of 3.3 liters and delivered 55 kW (75 hp) at 3000 rpm. It used two valves per cylinder and an overhead camshaft with upright shaft drive. The Auto Union designers had also equipped the engine with a Solex register carburetor. The four-gear transmission of the 920 transferred the power to the rear wheels.

The car could achieve a top speed of 130 km/h, but only temporarily. On the new autobahns of the Reich, it transpired that many engines of the time were not suitable for a longer ride at full throttle. Thus in 1938, the Auto Union specified the "Autobahn cruising speed" as 118 km/h only.

Der 3,3-Liter-Sechszylinder-Reihenmotor des Audi 920 leistete 55 kW (75 PS) bei 3.000 Umdrehungen.

The 3.3-liter 6-cylinder in-line engine of the Audi 920 provided 55 kW (75 hp) at 3000 rpm.

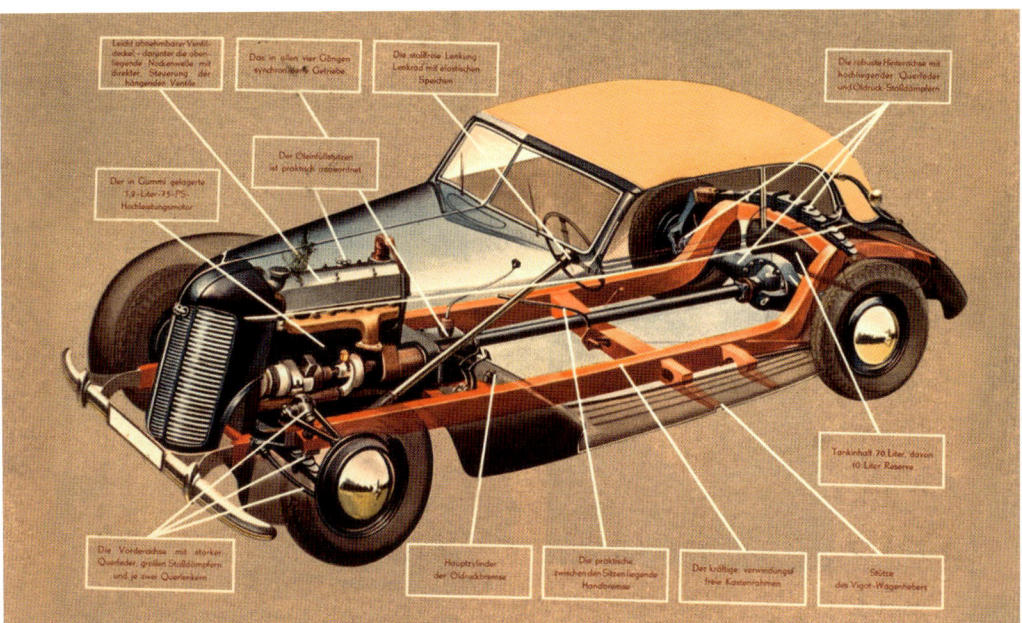

Aufgeschnitten: So zeigte Audi das neue Modell 920 im Verkaufsprospekt von 1938.

Cutaway: The new Audio model 920 in an advertising brochure from the year 1938.

Die Marke Audi präsentierte mit dem 920 im Jahr 1938 ein Modell der gehobenen Mittelklasse. Der Wagen war ein Produkt des Baukastensystems der Auto Union – so basierte der Motor auf dem Achtzylinder von Horch.

In 1938, Audi presented the 920 as a model of the upper medium class. The car was a product of the modular construction system of the Auto Union, e.g. the engine was based on the 8-cylinder engine by Horch.

Der Auto Union Typ D mit Doppelkompressor leistete 356 kW (485 PS). Der Rennwagen wog 850 Kilogramm.

The Auto Union type D with double compressor provided 356 kW (485 hp) and weighed 850 kg.

Modern: Instrumente aus der Luftfahrt lieferten den Auto-Union-Piloten die wichtigsten Informationen.

Modern cockpit: Aircraft instrumentation gave the Auto Union drivers all the important information.

Auto Union „Silberpfeil" Typ D

Fahrzeugdaten

Hersteller:	Auto Union
Land:	Deutschland
Modell:	Typ D
Bauzeit:	1938–1939
Länge:	4.200 mm
Breite:	1.660 mm
Höhe:	1.060 mm
Radstand:	2.800 mm
Leergewicht:	850 kg
Antriebsart:	Hinterrad
Höchstgeschwindigkeit:	ca. 340 km/h
Verbrauch:	k.A.

*Werte für 1939 Auto Union Typ D

Seit 1934 lieferten sich die Auto Union und Mercedes-Benz mit ihren Rennwagen auf den Grand-Prix-Rennstrecken harte Duelle. Technisch unterschieden sich die „Silberpfeile" aus Zwickau und Untertürkheim deutlich: Während die Ingenieure von Mercedes-Benz auf das traditionelle Konzept des Frontmotors setzten, platzierten die Techniker der Auto Union das Antriebsaggregat als Mittelmotor hinter dem Fahrer. Diesem Konzept blieben die Sachsen auch beim Typ D treu.

Das Modell feierte sein Debüt 1938 in einer schwierigen Zeit für die Auto Union, denn nur kurz zuvor war Bernd Rosemeyer tödlich verunglückt. Doch die Verantwortlichen fanden mit Tazio Nuvolari einen Ersatz für ihren Star. Nuvolari bescherte der Auto Union 1938 immerhin zwei Siege in Monza und Donington Park. Auch 1939 gab es Triumphe für den Typ D: Hermann Paul Müller gewann in Reims, während Tazio Nuvolari das letzte Rennen dieser Ära am 3. September 1939 in Belgrad für sich entschied.

Specifications

Manufacturer:	Auto Union
Country:	Germany
Model:	Type D
Produced:	1938–1939
Length:	4200 mm
Width:	1660 mm
Height:	1060 mm
Wheelbase:	2800 mm
Empty weight:	850 kg
Type of drive:	Rear-wheel drive
Max. speed:	approx. 340 km/h
Fuel consumption:	n.a.

* data for the 1939 Auto Union type D

Since 1934, Auto Union and Mercedes-Benz fought fierce duels with their Grand Prix racing cars. There were significant technological differences between the "silver arrows" from Zwickau and Untertürkheim: while the Mercedes-Benz engineers relied on the traditional concept of a front engine, the Auto Union technicians placed the engine behind the driver as a so-called midship engine. The Saxon manufacturer also used this concept for the type D.

This model debuted in 1938. These were bad days for Auto Union, as Bernd Rosemeyer had been killed in a fatal accident only a short time ago. However, the managers were able to find a replacement for their star, namely Tazio Nuvolari. In 1938, he achieved two wins for Auto Union in Monza and Donington Park. The type D could also triumph in 1939; Hermann Paul Müller won in Reims, and Tazio Nuvolari won the last race of the era in Belgrade on September 3rd 1939.

Der Motor

Motordaten

Bauart:	Viertakt-60-Grad-V-Motor
	mit Kompressor
Zylinderzahl:	12
Hubraum:	2.990 cm³
Bohrung:	65 mm
Hub:	75 mm
Leistung:	356 kW (485 PS)
	bei 7.000/min

Ab 1938 führte die internationale Sporthoheit AIACR, Vorläufer der heutigen FIA, ein neues technisches Reglement für Grand-Prix-Rennwagen ein. Kompressor-Triebwerke durften nur noch über einen Hubraum von maximal 3,0 Liter verfügen, während Saugmotoren 4,5 Liter Hubraum haben durften. Die Auto Union setzte beim Antriebskonzept für die Neukonstruktion Typ D weiter auf einen aufgeladenen Motor, der mit einem Roots-Kompressor ausgestattet war. Doch während das Antriebsaggregat im Vorgängermodell Typ C noch 16 Zylinder besessen hatte, entwickelten die Ingenieure der Auto Union für den neuen Rennwagen einen V12-Antrieb. Beim Renndebüt 1938 leistete das Triebwerk im neuen Grand-Prix-Rennwagen 308 kW (420 PS). Ein Jahr später kam eine überarbeitete Version des Rennmotors zum Einsatz: Ein Doppelkompressor sorgte nun für 356 kW (485 PS) bei 7.000 Umdrehungen pro Minute. Damit verfügte der Typ D über eine ähnliche PS-Leistung wie der Mercedes-Benz W154.

The engine

Engine specifications

Type:	Four-stroke 60° V engine
	with compressor
Number of cylinders:	12
Displacement:	2990 cm³
Bore:	65 mm
Stroke:	75 mm
Power:	356 kW (485 hp)
	at 7000 rpm

From 1938 on, the international sports organization AIACR, predecessor of the FIA of today, introduced new technological rules for Grand Prix racing cars. The displacement of compressor engines was limited to 3.0 liters while suction engines were allowed a displacement of 4.5 liters. For the drive concept of the newly designed type D, Auto Union still relied on a charged engine with a Roots compressor. While the engine of the predecessor model type C had 16 cylinders, the Auto Union engineers developed a V12 engine for the new racing car. At its racing debut in 1938, the engine of the new Grand Prix racing car provided 308 kW (420 hp). In the next year, an advanced version of the racing engine was used. A double compressor now provided for 356 kW (485 hp) at 7000 rpm. Thus the type D has a similar engine power as the Mercedes-Benz W154.

Die Einzelauspuffrohre waren extrem kurz, der daraus ertönende Klang war umso gewaltiger.

The single exhaust pipes were extremely short, and the ensuing sound was thus tremendous.

Monument: Der zweistufige Kompressor, rechts daneben die Vergaser-Batterie.
Monumental: The two-stage compressor with the carburetor battery on the right.

Leistungsfähig: Die Kraftstoffpumpen des V12-Motors stammten aus dem Flugzeugbau.
Powerful: The fuel pumps of the V12 engine derived from aircraft engineering.

Der Typ D war der letzte Grand-Prix-Rennwagen, den die Auto Union auf die Räder stellte. In seiner letzten Version mit Doppelkompressor erreichte der V12 eine Leistung von 356 kW (485 PS) bei 7.000 Umdrehungen.

The type D was the last Grand Prix racing car built by Auto Union. The last version of the V12 engine with twin compressor achieved a power of 356 kW (485 hp) at 7000 rpm.

Das elegante viersitzige DKW F8 „Front Luxus Cabriolet" wurde nur von 1939 bis 1940 gebaut.
The elegant four-seater DKW F8 "Front Luxus Cabriolet" was produced only from 1939 to 1940.

DKW F8

Fahrzeugdaten

Hersteller:	DKW
Land:	Deutschland
Modell:	F8 Front Luxus Cabriolet
Bauzeit:	1939–1940
Länge:	4.000 mm
Breite:	1.490 mm
Höhe:	1.480 mm
Radstand:	2.600 mm
Leergewicht:	790–830 kg
Antriebsart:	Vorderrad
Höchstgeschwindigkeit:	85 km/h
Verbrauch:	7 Liter/100 km

Die Anfänge von DKW gehen auf das Jahr 1904 und die Firma Rasmussen & Ernst zurück. Während des Ersten Weltkriegs begann man mit der Entwicklung eines Dampfkraftwagens, der es aber nie in die Serienproduktion schaffte. Daraus hervor ging jedoch die Firmenbezeichnung DKW.

Der F8 wurde von 1939 bis 1942 in den drei Varianten „Reichsklasse", „Meisterklasse" und „Front Luxus Cabriolet" in rund 50.000 Einheiten produziert. Nach dem Zweiten Weltkrieg übernahm der Industrieverband Fahrzeugbau (IFA) in der DDR das Audi-Werk in Zwickau und ließ das Vorkriegsmodell als IFA F8 in kaum veränderter Form wieder aufleben.

Die DKW-F8-Karosserie aus kunstlederbespanntem Sperrholz war auf einen Stahl-Kastenrahmen aufgesetzt. Zu den weiteren Besonderheiten des F8 zählte der Frontantrieb, der für problemlose Fahreigenschaften sorgte. Das elegante Fahrzeug prägte noch lange nach seinem Produktionsende das deutsche Straßenbild und zählt heute zu den begehrten Sammlerobjekten.

Specifications

Manufacturer:	DKW
Country:	Germany
Model:	F8 Front Luxus Cabriolet
Produced:	1939–1940
Length:	4000 mm
Width:	1490 mm
Height:	1480 mm
Wheelbase:	2600 mm
Empty weight:	790–830 kg
Type of drive:	Front-wheel drive
Max. speed:	85 km/h
Fuel consumption:	7 l/100 km

The history of DKW goes back to the year 1904 and the Rasmussen & Ernst company. During WWI, the company started to develop a steam car, which never made it into mass production but inspired the company name DKW ("Dampf-Kraft-Wagen," "steam motor vehicle"). 50,000 units of the F8 were built from 1939 to 1942. There were three models, "Reichsklasse" ("empire class"), "Meisterklasse" ("master class") and "Front Luxus Cabriolet." After WWII, the Industrial Association for Vehicle Construction (Industrieverband Fahrzeugbau, IFA) in the GDR took over the Audi factory in Zwickau and revived the pre-war model with only a few modifications as IFA F8.

The F8 body made of plywood covered with synthetic leather was mounted on a steel box frame. One of the few special features of the F8 was the front-wheel drive, which facilitated a trouble-free ride. Even long after the end of production, the elegant F8 still dominated the German streets. Today, it is a much sought-after collector's item.

Der Motor

Motordaten

Bauart:	Zweitakt-Reihenmotor
Zylinderzahl:	2
Hubraum:	692 cm³
Bohrung:	76 mm
Hub:	76 mm
Leistung:	14,7 kW (20 PS)
	bei 3.500/min

Mit der großen Zeit des deutschen Automobilbaus der 1930er-Jahre verbindet der Fan vor allem großvolumige Motoren. Daneben gab es aber auch kleine Aggregate, wie den Motor des DKW F8. DKW griff auf ein bewährtes Motorenkonzept zurück, das 1932 bereits im F2 zum Einsatz gekommen war: ein quer eingebauter Zweizylinder-Zweitakt-Reihenmotor. Im Basismodell des F8, der „Reichsklasse", arbeitete der kleinere 0,6-Liter-Motor, der es auf eine Leistung von 13,2 kW (18 PS) brachte. Für die „Meisterklasse" und das „Front Luxus Cabriolet" gab es eine Variante mit 692 cm³ Hubraum und 14,7 kW (20 PS) bei 3.500 Umdrehungen pro Minute. Typisch für diese DKW-Motoren war die Schnürle-Umkehrspülung. Mit dieser Art des Ladungswechsels im Zylinder eines Zweitakters konnte die Motorleistung gesteigert und gleichzeitig der spezifische Verbrauch gesenkt werden. Laut Angaben begnügte sich der F8 daher mit nur 7 Liter Kraftstoff auf 100 Kilometer.

The engine

Engine specifications

Type:	Two-stroke in-line engine
Number of cylinders:	2
Displacement:	692 cm³
Bore:	76 mm
Stroke:	76 mm
Engine power:	14.7 kW (20 hp)
	at 3500 rpm

The heydays of German car production in the 1930s are particularly associated with large-displacement engines. In addition, there were also smaller engines, like the one of the DKW F8. With this engine, DKW harked back to a successful concept that was already used in the F2 in 1932: a two-cylinder two-stroke in-line engine mounted crosswise. The F8 standard model "Reichsklasse" used a smaller 0.6 l engine with an engine power of 13.2 kW (18 hp). The "Meisterklasse" and "Front Luxus Cabriolet" models had an engine with a displacement of 692 cm³ and an engine power of 14.7 kW (20 hp) at 3500 rpm. A typical feature of these DKW engines was the Schnuerle porting. This type of scavenging made it possible to increase the engine power and at the same time decrease the fuel consumption. According to specifications, the F8 thus contented itself with just 7 liters of fuel for 100 kilometers.

Der DKW F8 verfügte über einen Stahl-Kastenrahmen, Einzelradaufhängung vorn und Frontantrieb.

The DKW F8 had a steel box frame, independent wheel suspension in the front and a front-wheel drive.

Bauartbedingt geriet der Zweizylinder-Zweitaktmotor des DKW F8 klein, kompakt und dennoch kräftig.

The two-cylinder two-stroke engine of the DKW F8 was designed to be small and compact and yet powerful.

In DKW-Tradition sorgte im F8 ein knapp 700 cm³ großer Zweitaktmotor für Vortrieb. Mit 14,7 kW (20 PS) Leistung ermöglichte er dem leichten Wagen immerhin 85 km/h Höchstgeschwindigkeit.

According to the grand tradition of DKW, the F8 was driven by a two-stroke engine with just under 700 cm³. With its engine power of 14.7 kW (20 PS), it provided the light car with a maximum speed of as much as 85 km/h.

Vom TPV wurden 250 Prototypen gebaut. Der Zweite Weltkrieg verhinderte eine Serienfertigung.
250 prototypes of the TPV were built. The outbreak of World War II prevented mass production.

Citroën TPV

Fahrzeugdaten

Hersteller:	Citroën
Land:	Frankreich
Modell:	TPV
Bauzeit:	1938–1939
Länge:	3.780 mm
Breite:	1.480 mm
Höhe:	1.600 mm
Radstand:	2.350 mm
Leergewicht:	380 kg
Antriebsart:	Vorderrad
Höchstgeschwindigkeit:	60 km/h
Verbrauch:	3 Liter/100 km

Der Citroën 2CV, auch liebevoll „Ente" genannt, ist vielen Automobil-Fans in bester Erinnerung. Weniger bekannt ist der Vorgänger der „Ente" – der TPV. Die drei Buchstaben stehen für „toute petite voiture", was so viel wie „ganz kleines Auto" heißt. Ein solches gab Citroën-Chef Pierre-Jules Boulanger 1934 in Auftrag. Der minimalistische Kleinwagen sollte Platz für zwei Bauern und einen Zentner Kartoffel bieten, mindestens 60 km/h erzielen und dabei nur 3 Liter Benzin pro 100 Kilometer konsumieren. Weiter sollte der Wagen robust sein und auch auf schlechten Wegen ein Mindestmaß an Komfort bieten. Bis 1939 wurden vom TPV – auch „2CV Serie A" genannt – 250 Prototypen gebaut. Schönheit spielte dabei keine Rolle, und so besaß der Wagen nur das, was er unbedingt brauchte. So begnügte man sich mit nur einem Scheinwerfer, auf ein zweites Rücklicht wurde ebenfalls verzichtet. Fortschrittlich war der Einsatz von Leichtmetall, was den Wagen anfangs nur 380 Kilogramm schwer geraten ließ.

Specifications

Manufacturer:	Citroën
Country:	France
Model:	TPV
Produced:	1938–1939
Length:	3780 mm
Width:	1480 mm
Height:	1600 mm
Wheelbase:	2350 mm
Empty weight:	380 kg
Type of drive:	Front-wheel drive
Max. speed:	60 km/h
Fuel consumption:	3 l/100 km

Many car fans still have fond memories of the Citroën 2CV, also called the "ugly duckling." Less well known is the predecessor of the "duckling," namely the TPV. The three letters stand for "toute petite voiture," meaning "very small car." Such a car was commissioned by Citroën director Pierre-Jules Boulanger in 1934. The minimalistic subcompact car should accommodate two peasants and a hundredweight of potatoes, achieve a speed of at least 60 km/h and consume no more than 3 liters of gasoline per 100 km. Additionally, the car should be robust and provide a minimum of comfort even on rough tracks. 250 prototypes of the TPV, also called "2CV Series A," were built until 1939. Beauty was deemed irrelevant, and hence the car only got what it absolutely required. The designers made do with only one headlight and even dispensed with a second backlight. The use of light metal, on the other hand, was quite progressive and made the car weigh no more than 380 kg at first.

Der Motor

Motordaten

Bauart:	Viertakt-Boxer-Motor
Zylinderzahl:	2
Hubraum:	375 cm³
Bohrung:	62 mm
Hub:	62 mm
Leistung:	5,9 kW (8 PS)
	bei 3.200/min

Dem Konzept des Citroën TPV entsprechend, fand sich in ihm nur ein kleiner, kompakter Motor. Er hatte gerade einmal einen Hubraum von 375 cm³ und lieferte bescheidene 5,9 kW (8 PS). Dabei handelte es sich um einen Zweizylinder-Boxermotor, der im Gegensatz zum späteren 2CV-Aggregat jedoch noch wasser- und nicht luftgekühlt war. Der Boxermotor hatte bis dahin vor allem in BMW- Motorrädern sowie in den Wagen von Boxer-Erfinder Carl Benz Karriere gemacht. Etwa zur selben Zeit baute auch Ferdinand Porsche in seinen KdF-Wagen (Vorläufer des VW Käfer) einen Boxermotor ein und läutete damit die Erfolgsgeschichte dieses Motorkonzepts im Automobilbau ein.

Einen Anlasser hatte man dem Motor des Citroën TPV nicht gegönnt. Stattdessen gab es nur eine Anlasserkurbel, die nach Ansicht der Citroën-Chefetage die Frauen der Bauern, für die der Wagen gedacht war, betätigen könnten. Immerhin stattete man den Kleinwagen mit einem Dreigang-Schaltgetriebe und Frontantrieb aus.

The engine

Engine specifications

Type:	4-stroke flat engine
Number of cylinders:	2
Displacement:	375 cm³
Bore:	62 mm
Stroke:	62 mm
Power:	5.9 kW (8 hp)
	at 3200/min

Matching its general concept, the Citroën TPV only had a small and compact engine with a displacement of just 375 cm³ and a humble power of 5.9 kW (8hp). In contrast to the later engine of the 2CV, this 2-cylinder flat engine still used water cooling instead of air cooling. Up to this time, the flat engine had mostly been used in BMW motorcycles and in the cars of its inventor Carl Benz. Nearly at the same time, Ferdinand Porsche mounted a flat engine in his KdF car (the predecessor of the Volkswagen Beetle) and thus started the success story of this engine concept in the automobile industry. The engine of the Citroën TPV had not been granted a starter; instead, there was only a crank. The upper management of Citroën was of the opinion that the wives of the peasants for whom the car was designed should handle the crank. At least the subcompact car was equipped with a manual three-gear transmission and a front-wheel engine.

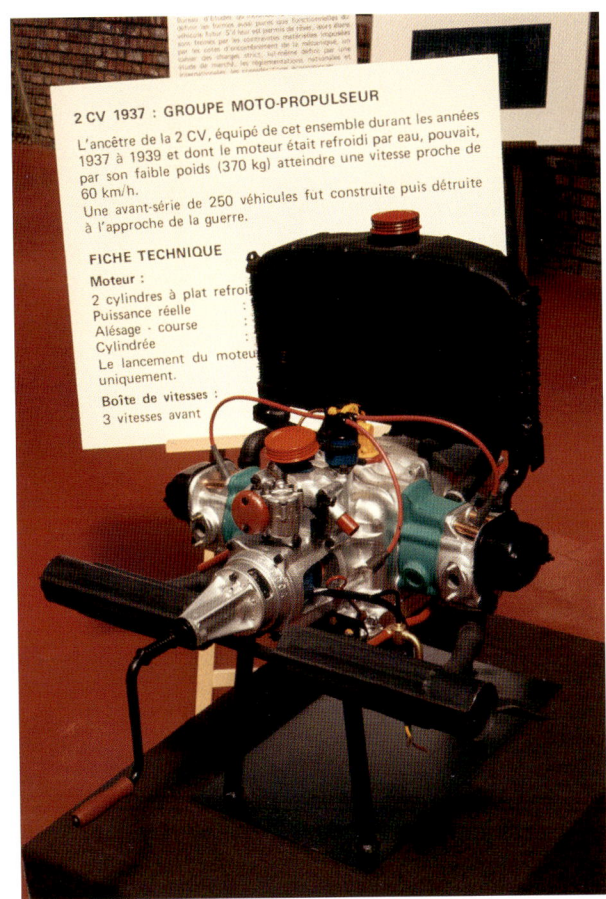

Der Boxermotor des Citroën TPV war im Gegensatz zum 2CV-Aggregat noch wassergekühlt.
In contrast to the engine of the 2CV, the flat engine of the Citroën TVP still used water cooling.

Der Citroën TPV war der unmittelbare Vorgänger der späteren, weltberühmten „Ente". Der Zweizylinder-Boxermotor des TPV war jedoch noch wassergekühlt.

The Citroën TPV was the immediate predecessor of the world-famous "ugly duckling." However, the 2-cylinder flat engine of the TPV still used water cooling.

Der Willys MB, hier von 1943, wurde allein während des Zweiten Weltkriegs an die 640.000-mal gebaut.

During WWII alone, approx. 640,000 units of the Willys MB were built (here a model from the year 1943).

Willys MB

Fahrzeugdaten

Hersteller:	Willys-Overland Company
Land:	USA
Modell:	Willys MB
Bauzeit:	1941–1945
Länge:	3.327 mm
Breite:	1.575 mm
Höhe:	1.829 mm
Radstand:	2.032 mm
Leergewicht:	1.113 kg
Antriebsart:	Allrad
Höchstgeschwindigkeit:	ca. 100 km/h
Verbrauch:	10–20 Liter/100 km

Er ist vermutlich der bekannteste Wagen aus dem Zweiten Weltkrieg: der Willys MB, besser bekannt als Willys Jeep. Seine Entwicklung geht auf eine Ausschreibung der US-Streitkräfte von 1940 zurück, die einen geländegängigen und zuverlässigen Wagen suchten, der sich preiswert in großer Stückzahl herstellen ließ. Alleine von 1941 bis Kriegsende wurden an die 640.000 Einheiten gebaut.

Der Willys MB war an allen Fronten zu finden; eine Folge seiner sehr guten Geländegängigkeit und Anpassungsfähigkeit. Auf der Straße erreichte er fast 100 km/h, im Feld war für ihn dank seines zuschaltbaren Allradantriebs kein Hindernis zu schwer. Da man sich bei dem Geländewagen auf das Notwendige beschränkte, war er wartungsfreundlich und ließ sich schnell und leicht reparieren. Der bald nur noch „Jeep" genannte Wagen verrichtete bis weit in die 1960er-Jahre bei vielen Streitkräften seinen Dienst. Zudem fand er zahllose zivile Einsatzmöglichkeiten.

Specifications

Manufacturer:	Willys-Overland Company
Country:	USA
Model:	Willys MB
Produced:	1941–1945
Length:	3327 mm
Width:	1575 mm
Height:	1829 mm
Wheelbase:	2032 mm
Empty weight:	1113 kg
Type of drive:	All-wheel drive
Max. speed:	approx. 100 km/h
Fuel consumption:	10–20 l/100 km

The Willys MB, better known as Willys Jeep, is probably the most famous car of World War II. Its development harks back to 1940 when the US armed forces invited tenders for a reliable all-terrain car that could be built cheaply in large quantities. From 1941 until the end of the war alone, approx. 640,000 units were built.

Due to its very good all-terrain capabilities and its adaptability, the Willys MB was present in every war theater. On the road, it nearly achieved 100 km/h, and on rough terrain, no obstacle was too hard to navigate thanks to its switch-on all-terrain drive. As the equipment was limited to the essential, the car was highly maintainable and could be repaired quickly and easily. Until well into the 1960s, the car – soon solely called "Jeep" – served with many armed forces all around the world. In addition, there were also numerous civil applications.

Der Motor

Motordaten

Bauart:	Viertakt-Reihenmotor
Zylinderzahl:	4
Hubraum:	2.199 cm³
Bohrung:	79,38 mm
Hub:	111,13 mm
Leistung:	44 kW (55 PS)

Der Willys MB wurde von einem simpel aufgebauten, dafür umso robusteren, wassergekühlten Vierzylinder-Reihenmotor vom Typ L134 mit einem Hubraum von 2,2 Liter angetrieben. Er entfaltete seine maximale Leistung von etwa 40 kW bei 4.000 Umdrehungen pro Minute. Sein höchstes Drehmoment von 129 Nm gab er bereits bei 2.000 Touren ab. Das Verdichtungsverhältnis des Aggregats lag bei 6,5:1.

Jeder Zylinder war mit zwei seitlich angeordneten Ventilen bestückt. Dadurch wurde eine niedrige Bauhöhe des Zylinderkopfs erreicht, was den Motor insgesamt kompakt machte. Diese sogenannten SV-Motoren erkannte man daran, dass der Zylinderkopf keine Teile des Ventiltriebs enthielt. Im Normalbetrieb wirkte das Dreigang- und Reduktionsgetriebe auf die Hinterräder; der Allradantrieb war zuschaltbar.

Der Motor des Willys MB kam in modifizierter Form auch im ab 1949 gebauten Willys M38 zum Einsatz. Diese neuere Variante bot mit 44 kW (60 PS) eine höhere Leistung.

The engine

Engine specifications

Type:	4-stroke in-line engine
Number of cylinders:	4
Displacement:	2199 cm³
Bore:	79.38 mm
Stroke:	111.13 mm
Power:	44 kW (55 hp)

The Willys MB was driven by a simply constructed but all the more robust water-cooled 4-cylinder in-line engine type L134 with a displacement of 2.2 liters. It achieved its maximum power of approx. 40 kW at 4000 rpm. The maximum torque of 129 Nm was already achieved at 2000 rpm. The compression ratio amounted to 6.5:1.

Each of the cylinders was equipped with two side-mounted valves. This provided for a low height of the cylinder head and thus a compact shape of the engine. The so-called SV engines can be distinguished by the fact that the cylinder head does not contain any parts of the valve train. In normal operation, the 3-gear reduction transmission put the power to the rear wheels. The all-wheel drive could be switched on additionally.

A modified version of the Willys MB engine was used in the Willys M38, which was built beginning in 1949. This new version provided a higher engine power of 44 kW (60 hp).

Der Willys MB wurde von einem robusten 2,2-Liter-Reihenmotor angetrieben.

The Willys MB was driven by a robust 2.2 liter in-line engine.

Der erste Porsche: der 356/1 Roadster, der 1948 in Handarbeit in Gmünd/Kärnten gebaut wurde.
The first Porsche: the 356/1 Roadster handcrafted in Gmünd (Kärnten) in 1948.

Porsche 356

Fahrzeugdaten

Hersteller:	Porsche 356
Land:	Deutschland
Modell:	356
Baujahr:	1948–1965
Länge:	3.950 mm
Breite:	1670 mm
Höhe:	1.300 mm
Radstand:	2.100 mm
Leergewicht:	680 kg (Coupe)
Antriebsart:	Hinterrad
Höchstgeschwindigkeit:	140 km/h
Verbrauch:	k.A.

Nach dem Zweiten Weltkrieg war wegen der Zerstörungen und der Materialknappheit zunächst nicht an eine geregelte Autoproduktion in Europa zu denken. Im österreichischen Gmünd entstand zu jener Zeit in Handarbeit dennoch eine Legende: der erste Porsche 356.

Das Design stammte von Erwin Komenda, der schon dem VW Käfer zu seinem Aussehen verholfen hatte. Die Entwickler achteten darauf, möglichst viele Teile der gerade angelaufenen Käfer-Produktion zu übernehmen. So fand sich im 356er ein modifizierter Käfer-Motorblock; auch wurden Chassis, Lenkung, Getriebe, Radaufhängung und Bremsanlage übernommen. Eine Eigenentwicklung war indes der Gitterrohr-Rahmen mit Aluminium-Karosserie. Der Motor wurde vor der Hinterachse als Mittelmotor eingebaut, für die Serienfertigung wanderte er hinter die Achse.

Obwohl der erste 356er keinesfalls übermotorisiert war, überzeugte er aufgrund seiner Formgebung und seines geringen Gewichts mit überragenden Fahreigenschaften.

Specifications

Manufacturer:	Porsche 356
Country:	Germany
Model:	356
Produced:	1948–1965
Length:	3950 mm
Width:	1670 mm
Height:	1300 mm
Wheelbase:	2100 mm
Empty weight:	680 kg (coupe)
Type of drive:	Rear-wheel drive
Max. speed:	140 km/h
Fuel consumption:	n.a.

Due to the destruction and the lack of materials, regular car manufacturing was out of the question in post-war Europe. Nonetheless, a legend was born at this time, handcrafted in Gmünd in Austria: the first Porsche 356.

It was designed by Erwin Komenda who had already created the visual appearance of the VW Beetle, which had just gone into production. The developers were careful to use as many parts of the Beetle as possible. Hence, the 356 not only had a modified Beetle engine block, but also inherited the chassis, the steering, the transmission, the wheel suspension and the braking system of the Beetle. The lattice tube frame and the aluminum autobody, on the other hand, were new developments. The engine was put in front of the rear axle as a midship engine; however, in the mass-produced model it was moved behind the axle.

The first 356 was by no means overpowered, but its shape, its low weight and the superior ride characteristics were convincing.

Der Motor

Motordaten

Bauart:	Viertakt-Boxermotor
Zylinderzahl:	4
Hubraum:	1.086 cm³
Bohrung:	73,5 mm
Hub:	64 mm
Leistung:	29 kW (40 PS)

Als Antrieb des Ur-Porsches kam ein umgebauter Käfer-Boxermotor zum Einsatz. Das 1,1-Liter-Aggregat sorgte bei einer Drehzahl von 4.200 Umdrehungen pro Minute für kraftvolle 29 kW (40 PS). Sein maximales Drehmoment von 70 Nm entwickelte er bei 2.800 Umdrehungen. Im Gegensatz zum VW-Motor wurde der Hubraum des Porsche-Aggregats von 1.131 cm³ auf 1.086 cm³ verringert. Dies wurde bei gleichbleibendem Hub von 64 mm durch eine Reduzierung des Bohrungsdurchmessers von 75 auf 73,5 mm erreicht. Diese Maßnahme war erforderlich, um den Motor an die 1.100er-Rennklasse anzupassen.

Sowohl der VW Käfer als auch der Porsche 356 besaßen ein zweiteiliges Kurbelgehäuse, das aus Silumin- oder Elektron-Leichtmetallguss hergestellt wurde. Die Zylinder waren bei beiden aus Grauguss gefertigt. Anstatt auf die Muldenkolben des Käfers setzte man beim Ur-Porsche auf gewölbte Domkolben. Dadurch konnte die Verdichtung auf 7:1 gesteigert werden.

The engine

Engine specifications

Type:	4-stroke flat engine
Number of cylinders:	4
Displacement:	1086 cm³
Bore:	73.5 mm
Stroke:	64 mm
Power:	29 kW (40 hp)

A modified Beetle flat engine was used to drive the original Porsche. At 4200 rpm, the 1.1 liter engine provided powerful 29 kW (40 hp). The maximum torque of 70 Nm was achieved at 2800 rpm. In contrast to the VW engine, the displacement of the Porsche engine was reduced from 1131 cm³ to 1086 cm³ by reducing the bore from 75 to 73.5 mm while maintaining the same stroke. This was necessary to adjust the engine to the racing class 1100.

The VW Beetle as well as the Porsche 356 had a two-part crankcase made of Silumin or Elektron light metal casting. In both cars, the cylinders were made of gray cast iron. Instead of the concave pistons of the Beetle, the original Porsche had convex pistons. This allowed for increasing the compression ratio to 7:1.

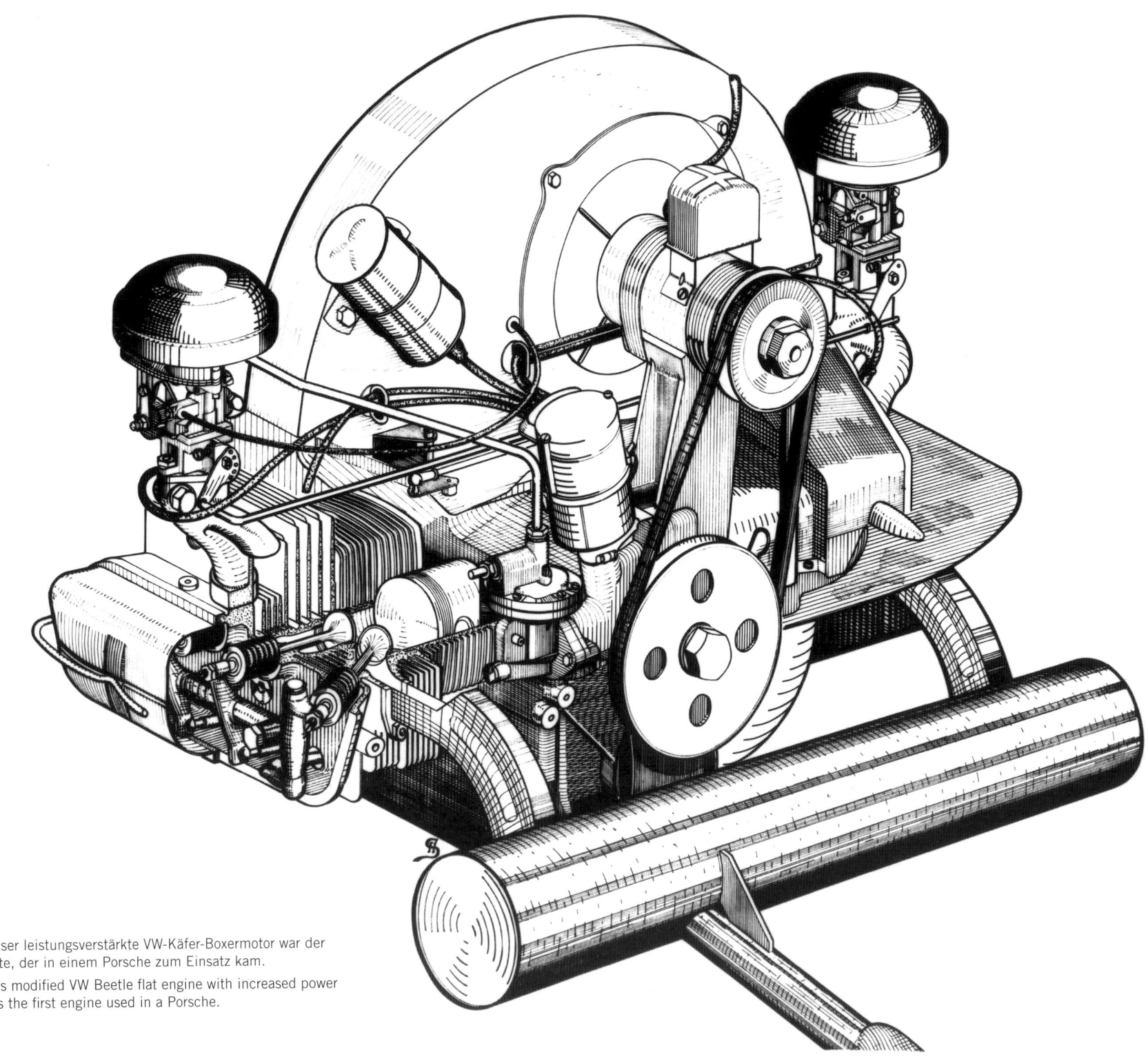

Dieser leistungsverstärkte VW-Käfer-Boxermotor war der
erste, der in einem Porsche zum Einsatz kam.

This modified VW Beetle flat engine with increased power
was the first engine used in a Porsche.

Mit dem Citroën 2CV (hier aus dem Jahr 1949) schuf Citroën weit mehr als nur ein Auto.
With the Citroën 2CV (here a 1949 model), Citroën created more than just a car.

Citroën 2CV „Ente" 1949

Fahrzeugdaten

Hersteller:	Citroën
Land:	Frankreich
Modell:	2CV
Bauzeit:	1949–1990
Länge:	3.778 mm
Breite:	1.478 mm
Höhe:	1.600 mm
Radstand:	2.370 mm
Leergewicht:	682 kg
Antriebsart:	Vorderrad
Höchstgeschwindigkeit:	64 km/h
Verbrauch:	7,8 Liter/100 km

*Werte für 2CV 1949

Am 7. Oktober 1948 wurde der 2CV auf dem Pariser Auto Salon vorgestellt. Zwar hatten die Grundzüge des Vorgängers TPV überlebt, doch mit dem 2CV kam nun ein stark überarbeitetes Modell. Die Karosserie bestand nicht mehr aus Aluminium-Wellblech, sondern aus Stahl. Auch einen zweiten Scheinwerfer sowie ein zweites Rücklicht hatte der 2CV spendiert bekommen, und die Innenausstattung wurde komfortabler. Im Sommer 1949 startete Citroën mit der Produktion; zuerst wurden Landwirte und Gewerbetreibende beliefert. Doch der 2CV bewährte sich nicht nur, er wurde auch rasch zum Kultfahrzeug. Die spartanische Ausstattung, der schwache Motor und das rückschiebbare Stoffdach, das im Sommer für Abkühlung sorgte, machten ihn zum Liebling. Der 2CV war nicht nur ein Auto – er verkörperte eine Lebenseinstellung. In Frankreich lief der letzte 2CV 1988 vom Band, und 1990 verließ das letzte Modell auch das portugiesische Citroën-Werk. Bis dahin waren rund 3,9 Millionen Einheiten gebaut worden.

Specifications

Manufacturer:	Citroën
Country:	France
Model:	2CV
Produced:	1949–1990
Length:	3778 mm
Width:	1478 mm
Height:	1600 mm
Wheelbase:	2370 mm
Empty weight:	682 kg
Type of drive:	Front-wheel drive
Max. speed:	64 km/h
Fuel consumption:	7.8 l/100 km

*Figures for 2CV 1949

On October 7th 1948, the 2CV was introduced at the Paris Auto Salon. The main features of the predecessor TPV had survived but the 2CV was a heavily reworked model. The auto body no longer consisted of corrugated aluminum sheet but of steel. The 2CV had also been treated to a second headlight and a second backlight, and the interior was more comfortable. Production started in summer 1949. Peasants and businesspeople were served first. The 2CV did not only prove its worth, it also became a cult vehicle. The spartan interior, the weak engine and the fabric roof, which provided relief from the heat in summer, made the car a firm favorite. The 2CV was not just a car but the epitome of a certain lifestyle. In France, production ended in 1988, and in 1990 the last 2CV left the Portuguese Citroën factory. Up to this time, approx. 3.9 million units had been built.

Der Motor

Motordaten

Bauart:	Viertakt-Boxer-Motor
Zylinderzahl:	2
Hubraum:	375 cm³
Bohrung:	62 mm
Hub:	62 mm
Leistung:	6,7 kW (9 PS)
	bei 3.800/min

*Werte für 2CV 1949

Bei der Motorisierung des ersten 2CV griff man im Wesentlichen auf den Zweizylinder-Boxermotor aus den Vorkriegs-Prototypen zurück. Dennoch gelang es, dem neuen Boxer mit 6,7 kW (9 PS) bei 3.800 Touren etwas mehr Leistung zu entlocken. Sein höchstes Drehmoment von 23 Nm entwickelte er dafür schon bei 1.800 Touren. Anstatt mit der Wasserkühlung der TPV-Prototypen ging der Motor mit Luftkühlung in Serie. Im Laufe der Jahrzehnte wurde auch der Boxermotor des 2CV weiterentwickelt. Er wurde etwas größer und gewann auch an Leistung. 1955 wurde der Hubraum auf 425 cm³ erweitert, was eine Leistungssteigerung auf 8,8 kW (12 PS) mit sich brachte. 1963 ließen sich dem kleinen Boxer 12 kW (16 PS) entlocken, und 1968 war die „Ente" auch mit einem 602-cm³-Motor mit 21 kW (28 PS) erhältlich. Gerade dieser Motor erwies sich als echter Langläufer und schaffte ohne Weiteres 300.000 Kilometer. Bis zuletzt ließ sich der kleine Boxermotor auch per Wagenheberkurbel starten.

The engine

Engine specifications

Type:	4-stroke flat engine
Number of cylinders:	2
Displacement:	375 cm³
Bore:	62 mm
Stroke:	62 mm
Power:	6,7 kW (9 hp) at 3800/min

* data for the 1949 2CV

For the engine of the first 2CV, Citroën basically harked back to the 2-cylinder flat engine of the pre-war prototypes. However, the engineers managed to extract slightly more power from the new flat engine, namely 6.7 kW (9 hp) at 3800 rpm. Instead of with the water cooling of the TPV prototypes, this engine was mass-produced with air cooling. During the decades of production, the flat engine of the 2CV was developed further. It became bigger and more powerful. In 1955, the displacement was extended to 425 cm³, which resulted in a power increase to 8.8 kW (12 hp). In 1963, the small flat engine provided 12 kW (16 hp), and in 1968, the duckling was also available with a 602 cm³ engine with 21 kW (28 hp). This engine proved to be really long-lived and could last for 300,000 km without difficulty. Until the end, the small flat engine could also be started by means of a car jack crank.

Im James-Bond-Streifen „In tödlicher Mission" von 1981 fuhr 007-Agent Roger Moore einen 2CV.

In the 1981 James Bond movie "For Your Eyes Only", agent 007 a.k.a. Roger Moore drove a 2CV.

Die Gebläse-gekühlten Boxermotoren der „Ente" waren simple, aber durchaus robuste Gesellen.

The fan-cooled flat engines of the "duckling" were simple but robust fellows.

Mit dem 1948 vorgestellten 2CV schuf Citroën nicht lediglich ein preiswertes und sparsames Auto, sondern ein Fahrzeug, das als „Ente" für Generationen sogar zu einer Art Weltanschauung wurde.

With the 2CV introduced in 1948, Citroën did not simply create a cheap and economic car, but also a vehicle that, nicknamed the "duckling", became a philosophy of life for several generations.

VW T1 Samba Bus bei der Kitzbühler Alpenralley 2014.
VW T1 Samba Bus at the 2014 Kitzbühl Alpine Rally.

VW T1 Samba Bus.
VW T1 Samba Bus.

Das Kennzeichen des VW Busses der ersten Generation war die geteilte Windschutzscheibe.
The split windscreen was a distinguishing mark of the first generation of the VW Bus.

VW Typ 2 T1

1950

Fahrzeugdaten

Hersteller:	Volkswagen
Land:	Deutschland
Modell:	Typ 2 T1
Baujahr:	1950–1967
Länge:	4.280 mm
Breite:	1.750 mm
Höhe:	1.940 mm
Radstand:	2.400 mm
Leergewicht:	975 kg
Antriebsart:	Hinterrad
Höchstgeschwindigkeit:	85 km/h
Verbrauch:	9 Liter/100 km

Der VW T1, besser bekannt als VW Bus oder auch Bulli, war nach dem Käfer das zweite Fahrzeug von Volkswagen. Das „T" in der Typenbezeichnung stand für „Transporter". Der T1 kam zu einer Zeit auf den Markt, als das deutsche Wirtschaftswunder gerade in seinen Startlöchern stand. Entsprechend groß war der Bedarf nach einem preiswerten und zuverlässigen Kleintransporter. Dieses Segment deckte der von 1950 bis 1967 gefertigte T1 perfekt ab: Es gab ihn in zahllosen Ausführungen vom simplen Transporter bis hin zum komfortablen Kleinreisebus. In späteren Jahren wurde der VW-Bus auch als Wohnmobil genutzt. Bis Auslaufen der T1-Serie wurden rund 1,8 Millionen Exemplare gebaut.

Bei der Motorisierung setzte man auf den bewährten Käfer-Boxermotor in einer durch Porsche modifizierten Form. Durch die Anpassung der Getriebeübersetzung wurde etwa eine bessere Beschleunigung erreicht; gleichzeitig wurde die Höchstgeschwindigkeit auf rund 85 km/h reduziert.

Specifications

Manufacturer:	Volkswagen
Country:	Germany
Model:	Type 2 T1
Produced:	1950–1967
Length:	4280 mm
Width:	1750 mm
Height:	1940 mm
Wheelbase:	2400 mm
Empty weight:	975 kg
Type of drive:	Rear-wheel drive
Max. speed:	85 km/h
Fuel consumption:	9 liters/100 km

The VW T1, better known as the VW Bus or Camper or even the "Hippie van", was the second Volkswagen vehicle after the Beetle. The "T" in the type designation stood for "Transporter." The T1 was launched when the German Wirtschaftswunder ("enonomic miracle") started to get rolling. Hence, there was a huge demand for a cheap and reliable small transporter, and this demand was perfectly covered by the T1 manufactured from 1950 to 1967. There were numerous variants from simple transporters to comfortable coaches. In later years, the VW Bus was also used as a motorhome. Approx. 1.8 million units were built until the T1 series was discontinued. For the power unit, the proven Beetle flat engine in the modified Porsche version was chosen. Adjustments to the transmission allowed for a better acceleration; however, they also reduced the maximum speed to approx. 85 km/h.

Der Motor

Motordaten

Bauart:	Viertakt-Boxermotor
Zylinderzahl:	4
Hubraum:	1.131 cm³
Bohrung:	75 mm
Hub:	64 mm
Leistung:	18,4 kW (25 PS)

In der von 1950 bis 1954 gebauten T1-Variante war ein 1,1-Liter-Aggregat im Einsatz, das seine maximale Leistung von 18,4 kW (25 PS) bei 3.300 Umdrehungen pro Minute entfaltet. Das maximale Drehmoment von 67 Nm stellte die luftgekühlte Maschine bei 2.000 Touren zur Verfügung. Ab 1954 brachte VW für den T1 einen stärkeren 1,2-Liter-Motor mit 22 kW (30 PS) auf den Markt. 1960 wurde dieser weiter verbessert und erreichte nun 25 kW (34 PS). Von 1963 bis 1967 wurde der T1 mit einer noch größeren 1,5-Liter-Maschine mit 31 beziehungsweise 32,3 kW (42 und 44 PS) ausgeliefert.

Der Motor war beim T1, ebenso wie bei den Nachfolgemodellen T2 und T3, im Heck in einem Kasten eingebaut. Dadurch besaß der Wagen keine durchgehende Ladefläche. Erst ab dem seit 1990 erhältlichen T4 verabschiedete man sich von diesem Konzept und spendierte dem VW Bus einen Frontmotor und Vorderradantrieb.

The engine

Engine specifications

Type:	4-stroke flat engine
Number of cylinders:	4
Displacement:	1131 cm³
Bore:	75 mm
Stroke:	64 mm
Power:	18.4 kW (25 hp)

In the T1 variant built from 1950 to 1954, an air-cooled 1.1 liter engine was used. It provided its maximum power of 18.4 kW (25 hp) at 3300 rpm and its maximum torque of 67 Nm at 2000 rpm. Beginning with 1954, VW used a more powerful 1.2 liter engine with 22 kW (30 hp) for the T1. In 1960, the engine was further improved and now provided 25 kW (34 hp). From 1963 to 1967, the T1 was equipped with an even bigger 1.5 liter engine with 31 or 32.3 kW (42 or 44 hp).

In the T1 as well as in the successor models T2 and T3, the engine was located in a box at the rear. This meant that the car did not have a continuous load floor. This concept was abandoned no sooner than in 1990 when the T4 with a front engine and a front-wheel drive became available.

Im VW Bus T1 arbeitete eine modifizierte Variante des Käfer-Boxermotors.

The VW Bus T1 was driven by a modified version of the Beetle flat engine.

Eine Kugel zum Knutschen.
The lovable bubble car.

BMW Isetta 250 Standard im BMW Museum in München.
BMW Isetta 250 Standard in the Munich BMW Museum.

BMW Isetta

Fahrzeugdaten

Hersteller:	BMW
Land:	Deutschland
Modell:	Isetta 250
Baujahr:	1955–1962
Länge:	2.286 mm
Breite:	1.384 mm
Höhe:	1.337 mm
Radstand:	1.473 mm
Leergewicht:	353 kg
Antriebsart:	Hinterrad
Höchstgeschwindigkeit:	82 km/h
Verbrauch:	etwa 3,45 Liter/100 km (bei 57 km/h)

Bis in die Mitte der 1950er-Jahre besaßen nur wenige Leute einen Auto-Führerschein. Deshalb waren Kleinwagen gefragt, die mit Motorrad-Führerschein pilotiert werden durften. 1954 stellte der italienische Motorrad-Hersteller Iso Rivolta seine Isetta vor, die wegen ihrer nach vorne aufgehenden Tür an einen Kühlschrank auf vier Rädern erinnerte.

BMW steckte zu jener Zeit in einer veritablen Krise und entschloss sich, einen Lizenzbau mit BMW-Motorradmotor anzubieten. Die BMW Isetta wurde ab März 1955 für 2.850 DM angeboten. Bis 1962 wurden 161.728 Stück verkauft.

Die Isetta war in vielem einzigartig: Die Sitzbank bot Platz für zwei Erwachsene. Um bequemes Einsteigen zu gewährleisten, waren Lenkrad und Armaturenbrett an der nach vorne öffnenden Tür angebracht. Eine Heizung gab es zunächst nur als Sonderzubehör. Dafür gehörte das Faltdach zur Serienausstattung. Minimalismus war auch im Innenraum angesagt: Es dominierte bedruckte Pappe.

Specifications

Manufacturer:	BMW
Country:	Germany
Model:	Isetta 250
Produced:	1955–1962
Length:	2286 mm
Width:	1384 mm
Height:	1337 mm
Wheelbase:	1473 mm
Empty weight:	353 kg
Type of drive:	Rear-wheel drive
Max. speed:	82 km/h
Fuel consumption:	approx. 3.45 liters/100 km (at 57 km/h)

Until the mid-1950s, only few people had a driver's license for a car. Hence, there was a demand for sub-compact cars that you were allowed to drive with just a motorcycle license. In 1954, the Italian motorcycle manufacturer Iso Rivolta introduced its Isetta. Thanks to its door opening forward, it was a bit reminiscent of a fridge on wheels.

At this time, BMW was suffering a veritable crisis and decided to offer a license build with a BMW motorcycle engine. Beginning with March 1955, the BMW Isetta was offered at 2850 DM. Until 1962, 161,728 units were sold.

The Isetta was unique in many respects: The bench accommodated two adults. The steering wheel and the dashboard were fastened to the forward opening door so that it was possible to enter the car in a comfortable way. At first, a heating was only available as special accessory. On the other hand, the folding roof was in base. Minimalism also dominated the interior: printed cardboard prevailed.

Der Motor

Motordaten

Bauart:	Viertakt-Ottomotor
Zylinderzahl:	1
Hubraum:	247 cm³
Bohrung:	68 mm
Hub:	68 mm
Leistung:	8,8 kW (12 PS)

Für die BMW Isetta wurden zwei Motoren angeboten. Jener der Isetta 250 hatte einen Hubraum von 247 cm³. Der Einzylinder-Ottomotor war quer hinter der Sitzbank an der rechten Wagenseiten eingebaut. Links davon waren eine Einscheiben-Trockenkupplung und das Getriebe montiert. Die Isetta verfügte über einen Rückwärts- und vier nicht synchronisierte Vorwärtsgänge, die beim Schalten das Geben von Zwischengas erforderten.

Das über ein Gebläse luftgekühlte Aggregat besaß eine Druckumlaufschmierung. Es lieferte seine maximale Leistung von 8,8 kW bei 5.800 Umdrehungen pro Minute. Sein Verdichtungsverhältnis lag bei 6,8:1.

1956 wurde unter der Typenbezeichnung 300 eine Isetta mit einem 297-cm³-Einzylindermotor angeboten. Seine Bohrung und sein Hub maßen 72 und 73 mm. Er brachte es auf 9,6 kW (13 PS) bei 5.200 Umdrehungen pro Minute. Sein Verdichtungsverhältnis betrug 7,0:1.

The engine

Engine specifications

Type:	4-stroke Otto engine
Number of cylinders:	1
Displacement:	247 cm³
Bore:	68 mm
Stroke:	68 mm
Power:	8.8 kW (12 hp)

The BMW Isetta was on offer with two different engines. The one in the Isetta 250 had a displacement of 247 cm³. This 1-cylinder Otto engine was mounted at right angle at the right-hand side behind the bench. On the left, the single-disk dry clutch and the transmission were located. The Isetta had a reverse gear and four unsynchronized forward gears. This made double-clutching necessary while shifting.

The engine was equipped with a fan for air cooling and a forced feed lubrication. It provided its maximum power of 8.8 kW at 5800 rpm. The compression ratio amounted to 6.8:1.

In 1956, the Isetta 300 with a 297 cm³ 1-cylinder engine was pitched. With a bore of 72 mm and a stroke of 73 mm, it achieved a maximum power of 9.6 kW (13 hp) at 5200 rpm and a compression ratio of 7.0:1.

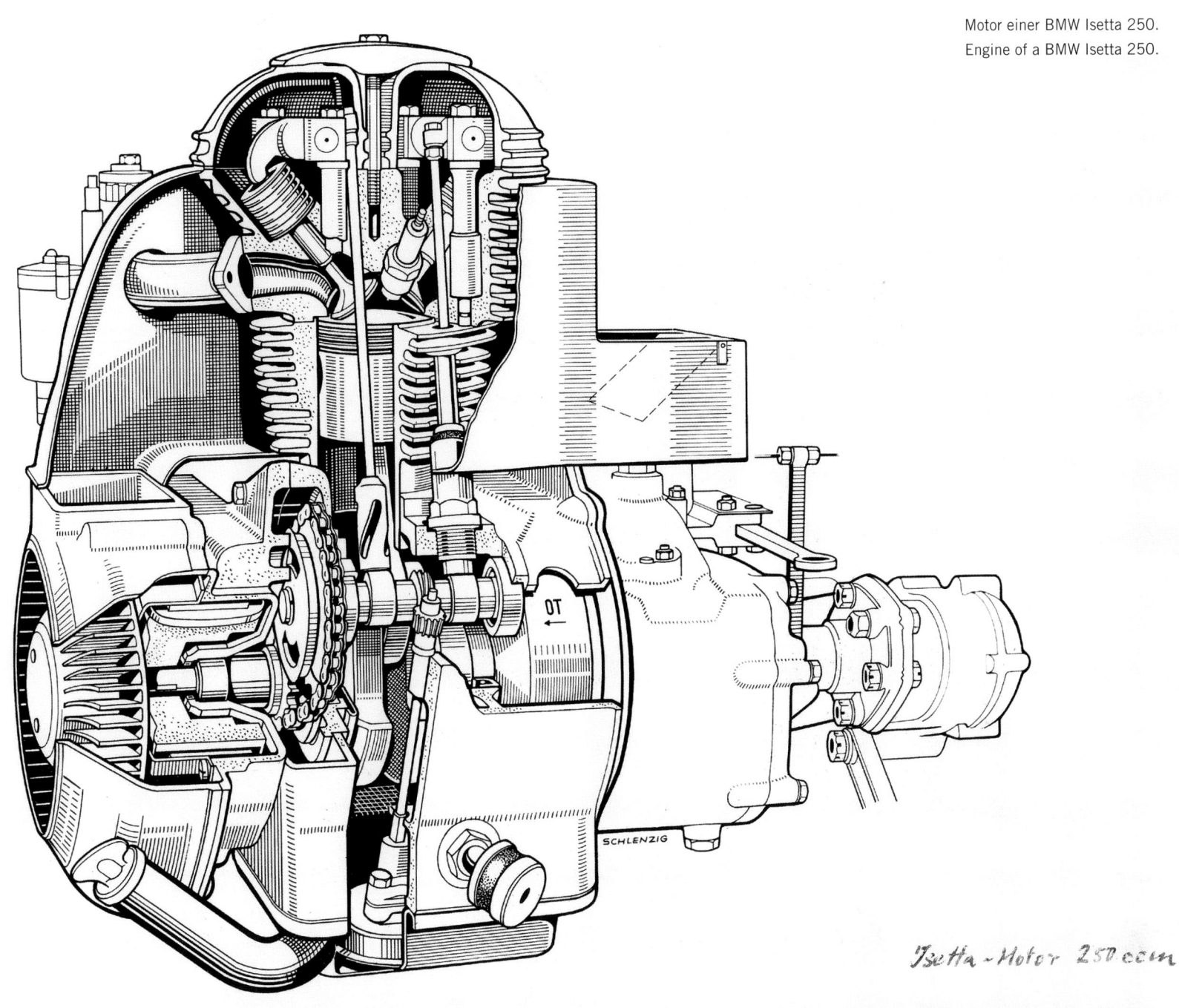

Motor einer BMW Isetta 250.
Engine of a BMW Isetta 250.

Isetta-Motor 250 ccm

Volvo Amazon 121.
Volvo Amazon 121.

Volvo Amazon

Fahrzeugdaten

Hersteller:	Volvo
Land:	Schweden
Modell:	Amazon
Baujahr:	1956–1970
Länge:	4.395 mm
Breite:	1.614 mm
Höhe:	1.510 mm
Radstand:	2.600 mm
Leergewicht:	1.130 kg
Antriebsart:	Hinterrad
Höchstgeschwindigkeit:	ca. 150 km/h
Verbrauch:	8–11 Liter/100 km

Der Volvo Amazon zählt zu den schönsten Oldtimern. Seine formschöne Karosserielinie vereinte die von amerikanischen Autos jener Tage bekannte Eleganz mit südländischem Flair. Markant waren ferner sein zweigeteilter, mit Chrom eingefasster Kühlergrill, die beiden Stoßstangenhörner an der Front und die stark gewölbte Heckscheibe.

Der im schwedischen Göteborg fabrizierte Wagen legte den Grundstein für den guten Ruf des Autobauers. So überzeugte der Amazon durch seine ausgezeichneten Fahrleistungen und seine hohe Sicherheit, die er bei Crashtests unter Beweis stellte. Ab 1959 war er weltweit der erste Wagen, der serienmäßig mit Dreipunkt-Sicherheitsgurten ausgestattet war.

Der Wagen führte nur in Skandinavien den Namen Amazon. Bei uns wurde er schlicht 120er-Volvo genannt, da sich der deutsche Motorradhersteller Kreidler den Namen Amazon bereits rechtlich gesichert hatte.

Specifications

Manufacturer:	Volvo
Country:	Sweden
Model:	Amazon
Produced:	1956–1970
Length:	4395 mm
Width:	1614 mm
Height:	1510 mm
Wheelbase:	2600 mm
Empty weight:	1130 kg
Type of drive:	Rear-wheel drive
Max. speed:	approx. 150 km/h
Fuel consumption:	8–11 liters/100 km

The Volvo Amazon is one of the most beautiful vintage cars. Its elegant shape combines the elegance of the American cars of its day with Mediterranean flair. Striking elements were the two-part radiator grille framed in chrome, the two bumper horns at the front and the significantly convex rear window.

The car was produced in Göteborg in Sweden and laid the foundation for the good reputation of its manufacturer. For example, the Amazon could score with its excellent driving characteristics and the high level of safety that had been proven in crash tests. As the first car in the world, it was equipped with three-point seat belts in base beginning with 1959.

The name Amazon was only used in Scandinavia. In other countries, it was simply called Volvo 120 because the German motorcycle manufacturer Kreidler had already reserved the rights for the name Amazon.

Der Motor

Motordaten

Bauart:	Viertakt-Otto-Reihenmotor
Zylinderzahl:	4
Hubraum:	1.582 cm³
Bohrung:	79,37 mm
Hub:	80 mm
Leistung:	44 kW (60 PS)

Der Volvo Amazon von 1956 war mit einem wassergekühlten Vierzylinder-Reihenmotor ausgestattet. Das 1,6-Liter-Aggregat entfaltete bei 4.500 Umdrehungen pro Minute seine höchste Leistung von 44 kW (60 PS). Sein maximales Drehmoment von 111 Nm stellte er bei 2.500 Umdrehungen pro Minute zur Verfügung. Das Verdichtungsverhältnis lag bei 7,5:1. Der Motor besaß eine unten liegende Nockenwelle sowie pro Zylinder zwei hängende Ventile, die über Stößelstangen und Kipphebel betätigt wurden. Die Kraft des Motors wurde über ein Dreigang-Schaltgetriebe auf die Hinterräder übertragen. Ab 1963 gab es das Fabrikat auch mit einem Automatik-Getriebe von Borg-Warner.

1961 wurde das 1,6-Liter-Aggregat von einer 1,8-Liter-Maschine mit 1.782 cm³ Hubraum abgelöst. Die Basisvariante brachte es auf 50 kW (68 PS). Weiterentwicklungen dieses Motors lieferten bis zu 71 kW (96 PS). In den letzten Amazon-Modellen ab 1968 bis 1970 versahen 2-Liter-Motoren mit 60 oder 76 kW ihren Dienst.

The engine

Engine specifications

Type:	4-stroke Otto in-line engine
Number of cylinders:	4
Displacement:	1582 cm³
Bore:	79.37 mm
Stroke:	80 mm
Power:	44 kW (60 hp)

The Volvo Amazon of 1956 was fitted with a water-cooled 4-cylinder in-line engine with a displacement of 1.6 liters. It provided its maximum power of 44 kW (60 hp) at 4500 rpm and its maximum torque of 111 Nm at 2500 rpm. The compression ratio amounted to 7.5:1. The engine had a camshaft in bottom position as well as two hanging valves per cylinder controlled by push rods and rocker levers. The engine power was transferred to the rear wheels via a three-gear manual transmission. Beginning with 1963, there was also an automatic transmission by Borg-Warner.

In 1961, the 1.6 liter engine was replaced by a 1.8 liter engine with a displacement of 1782 cm³. The standard version provided up to 50 kW (68 hp). Further advanced models of this engine even achieved up to 71 kW (96 hp). The last Amazon models from 1968 to 1970 were equipped with 2 liter engines with 60 or 76 kW.

Motor des Volvo Amazon.
Engine of the Volvo Amazon.

Das Bel Air Impala Hardtop (Sport)
Coupé von 1958 entstammte bereits
der zweiten Generation.

The Bel Air Impala coupé of 1958
is already a model of the second
generation.

Chevrolet Bel Air

Fahrzeugdaten

Hersteller:	Chevrolet
Land:	USA
Modell:	Bel Air
Bauzeit:	1954–1957
Länge:	5.080 mm
Breite:	1.877 mm
Höhe:	1.499 mm
Radstand:	2.921 mm
Leergewicht:	1.457 kg
Antriebsart:	Hinterrad
Höchstgeschwindigkeit:	159 km/h
Verbrauch:	ca. 25 Liter/100 km

*Werte für 1957 Hardtop-Coupé

Der Chevrolet Bel Air zählt heute zu den begehrten US-Klassikern. Insbesondere die Modelle der Baujahre 1953 bis 1957 sowie die Fahrzeuge der nur 1958 gefertigten zweiten Generation erfreuen in unseren Tagen die Herzen der Oldtimer-Fans. Typisch für die US-Cars der 1950er-Jahre war die extrovertierte optische Erscheinung, zu der die in Chrom schwelgende Front ebenso beitrug wie eine zweifarbige Lackierung, üppige Panoramascheiben und ausgeprägte Heckflossen. Das komfortabel abgestimmte Fahrwerk des mit einem Kastenrahmen ausgestatteten Mittelklassewagens setzte vorne auf eine Einzelradaufhängung mit Schraubenfedern, während achtern eine Starrachse mit Blattfedern zum Einsatz kam. Trotz vergleichsweise leistungsstarker Motorisierungen war der als zwei- oder viertürige Limousine, Hardtop (Sport) Coupé, Station Wagon, Townsman und Convertible erhältliche Bel Air nicht für sportliche Umtriebe, sondern für komfortables Reisen mit der ganzen Familie gedacht.

Specifications

Manufacturer:	Chevrolet
Country:	USA
Model:	Bel Air
Produced:	1954–1957
Length:	5080 mm
Width:	1877 mm
Height:	1499 mm
Wheelbase:	2921 mm
Empty weight:	1457 kg
Type of drive:	Rear-wheel drive
Max. speed:	159 km/h
Fuel consumption:	approx. 25 l/100 km

* data for the 1957 coupé

Today, the Chevrolet Bel Air is one of the most sought-after US classics. In particular the models of the years 1953 to 1958 and the vehicles of the second generation only produced in 1958 delight the fans of vintage cars. The extrovert visual appearance is typical for US cars of the 1950s. It shows in the front wallowing in chrome, the two-color paintwork, the extensive panorama windows and the distinctive tailfins. The comfortable undercarriage of the middle-class car with box frame had independent wheel suspension with spiral springs in the front and a fixed axle with leaf springs at the rear. The Bel Air was available as a two-door or four-door sedan, coupé, station wagon or convertible. Despite the relatively powerful engines, it was not designed for racing antics but for comfortable family rides.

Der Motor

Motordaten

Bauart:	Viertakt-90-Grad-V-Motor
Zylinderzahl:	8
Hubraum:	4.344 cm³
Bohrung:	95,25 mm
Hub:	76,2 mm
Leistung:	121 kW (165 PS) bei 4.400/min

Der Chevrolet Bel Air wurde sowohl mit Sechszylinder-Reihenmotoren als auch mit Achtzylinder-V-Motoren angeboten. Mit reichlich Drehmoment gesegnet, verhalf der 85 kW (115 PS) starke Reihensechszylinder dem 1.587 Kilogramm schweren Chevrolet Bel Air bereits zu guten Reisequalitäten. Dabei passte die ganz auf Durchzug ausgelegte Charakteristik des wohlklingenden Triebwerks perfekt zum Komfortanspruch des Bel Air und seiner Fahrgäste. Von 1955 an war der Bel Air auch mit dem legendären Small-Block-Triebwerk erhältlich. Dieser Achtzylinder-V-Motor mit 90 Grad Bankwinkel sollte gewissermaßen zum Synonym für Chevrolet schlechthin werden; das Triebwerk wurde in stetig weiterentwickelter Form bis heute annähernd 100 Millionen Mal gebaut. Aus 4.344 cm³ Hubraum entwickelte der V8 im Bel Air zu Anfang 121 kW (165 PS) bei 4.400 Umdrehungen pro Minute, während in der nächsten Generation 1958 bereits bis zu 232 kW (315 PS) aus 5.703 cm³ möglich waren.

The engine

Engine specifications

Type:	4-stroke 90° V-engine
Number of cylinders:	8
Displacement:	4344 cm³
Bore:	95.25 mm
Stroke:	76.2 mm
Engine power:	121 kW (165 hp) at 4400 rpm

The Chevrolet Bel Air was available with a 6-cylinder in-line engine as well as with an 8-cylinder V-engine. The high torque of the 85 kW (115 hp) 6-cylinder in-line engine already provided the 1587 kg Chevrolet Bel Air with good travel characteristics. The specifications of the melodious and powerful engine hence match the requirements for comfort of the Bel Air and its passengers. From 1955 on, the Bel Air was also available with the legendary Small Block engine. This 8-cylinder V-engine with a 90° bank angle would become sort of a synonym for Chevrolet. It was refined again and again, and up to now, approx. 100 million units have been built. With a displacement of 4344 cm³, the V8 in the Bel Air achieved 121 kW (165 hp) at 4400 rpm at first. In 1958, the second generation with a displacement of 5703 cm³ was already able to provide up to 232 kW (315 hp).

Seit seiner Einführung im Herbst 1954 wurde der V8-Small-Block annähernd 100 Millionen Mal gebaut.

Since its introduction in autumn 1954, approx. 100 million units of the V8 Small Block have been built.

Mit seiner geradezu barock anmutenden Formensprache verkörpert der mit reichlich Chrom versehene Chevrolet Bel Air den typischen amerikanischen Automobilbau der 1950er-Jahre.

With its nearly baroque design, the heavily chrome-laden Chevrolet Bel Air is the epitome of American automotive engineering in the 1950s.

Der Dodge Dart aus dem Jahr 1960 in der Ausstattungsvariante „Phoenix".
The 1960 Dodge Dart in the Phoenix version.

Dodge Dart

1960

Fahrzeugdaten

Hersteller:	Dodge
Land:	USA
Modell:	Dart
Bauzeit:	1960
Länge:	5.347 mm
Breite:	1.999 mm
Höhe:	1.392 mm
Radstand:	2.997 mm
Leergewicht:	1.750 kg
Antriebsart:	Hinterrad
Höchstgeschwindigkeit:	145 km/h
Verbrauch:	15–19 Liter/100 km

Als der US-Automobilhersteller Dodge zum Modelljahr 1960 den Dart präsentierte, stellte der als Limousine, Kombi, Coupé und Cabriolet erhältliche Wagen insbesondere ein erschwingliches Einstiegsmodell dar. Mit Ausnahme des Kombi besaß der in der Mittelklasse angesiedelte Dart deshalb im Vergleich zu den Modellen Matador und Polara einen um 10 Zentimeter verkürzten Radstand. Der äußerst gefällig gestaltete Wagen – im zeitgenössischen Stil mit reichlich Chrom und Heckflossen ausgestattet – wurde auf Anhieb zu einem vollen Erfolg. Allein 1960 verkaufte sich der Dart 306.603 Mal. Für das Modelljahr 1961 hatten die Dodge-Entwickler dem Dart ein umfassendes Facelift verpasst, dem die verspielten Verschnörkelungen an der Karosserie zum Opfer fielen. Ein eklatanter Fehlgriff der Designer, wie sich alsbald herausstellen sollte. Dass die Kunden mit dem neuen Dart nicht viel anzufangen wussten, zeigte sich an den um fast zwei Drittel eingebrochenen Verkaufszahlen.

Specifications

Manufacturer:	Dodge
Country:	USA
Model:	Dart
Produced:	1960
Length:	5347 mm
Width:	1999 mm
Height:	1392 mm
Wheelbase:	2997 mm
Empty weight:	1750 kg
Type of drive:	Rear-wheel drive
Max. speed:	145 km/h
Fuel consumption:	15–19 l/100 km

In 1960, the US car manufacturer Dodge presented the Dart, an affordable entry-level medium-class model available as sedan, station wagon, coupé and convertible. Except for the station wagon, the Dart had a wheelbase that was 10 cm shorter than that of the Matador and Polara models. The car's design was extraordinarily pleasing in the contemporary style with loads of chrome and huge tailfins, and the model became immediately a success. In 1960 alone, 306,603 units were sold. In the model year 1961, the Dodge developers gave the Dart an extensive facelift, which got rid of the playful flourishes. Very soon, however, this proved to be a huge mistake. The reduction of the sales figures by nearly two thirds showed that the customers could not warm to the new Dart.

Der Motor

Motordaten

Bauart:	Viertakt-Reihenmotor
Zylinderzahl:	6
Hubraum:	3.682 cm³
Bohrung:	86,4 mm
Hub:	104,8 mm
Leistung:	109 kW (147 PS)
	bei 4.000/min

In seiner Standardausführung wurde der in den drei Ausstattungsvarianten „Phoenix", „Pioneer" und „Seneca" angebotene Dodge Dart 1960 mit einem Reihen-Sechszylinder-Motor mit 109 kW (147 PS) Leistung bei 4.000 Umdrehungen pro Minute angeboten. Das maximale Drehmoment von 291 Nm lag bei 2.800 Umdrehungen pro Minute. Der durchzugskräftige und bauartbedingt sehr laufruhige Motor war mit jeweils zwei hängenden Ventilen pro Zylinder ausgestattet, die von einer seitlich angeordneten kettengetriebenen Nockenwelle gesteuert wurden. Die Gemischaufbereitung des wassergekühlten Triebwerks übernahm ein Fallstromvergaser. Auf Wunsch gab es den Dart auch mit einem kräftigen V8-Motor mit 5,2 Liter Hubraum und 171 kW (233 PS) sowie einer potenten 6,3-Liter-V8-Maschine mit 242 kW (329 PS). In der Grundausstattung war der Dart mit einem Dreigang-Schaltgetriebe ausgestattet. Als Sonderzubehör war er auch mit dem automatischen Torqueflite-Getriebe erhältlich.

The engine

Engine specifications

Type:	Four-stroke in-line engine
Number of cylinders:	6
Displacement:	3682 cm³
Bore:	86.4 mm
Stroke:	104.8 mm
Power:	109 kW (147 hp)
	at 4000 rpm

The Dodge Dart 1960 was available in the "Phoenix," "Pioneer" and "Seneca" versions. The basic configuration had a 6-cylinder in-line engine with 109 kW (147 hp) at 4000 rpm and a maximum torque of 291 Nm at 2800 rpm. The powerful engine had a smooth run by design and was equipped with two overhead valves per cylinder, which are controlled by a chain-driven camshaft at the side. Mixture control of the water-cooled engine was done by a down-draft carburetor. Optionally, the Dart was also available with a V8 engine with a displacement of 5.2 liters and 171 kW (233 hp) or with a 6.3-liter V8 engine with 242 kW (329 hp). The Dart was equipped with a manual three-gear transmission but an optional automatic Torqueflite transmission was also offered.

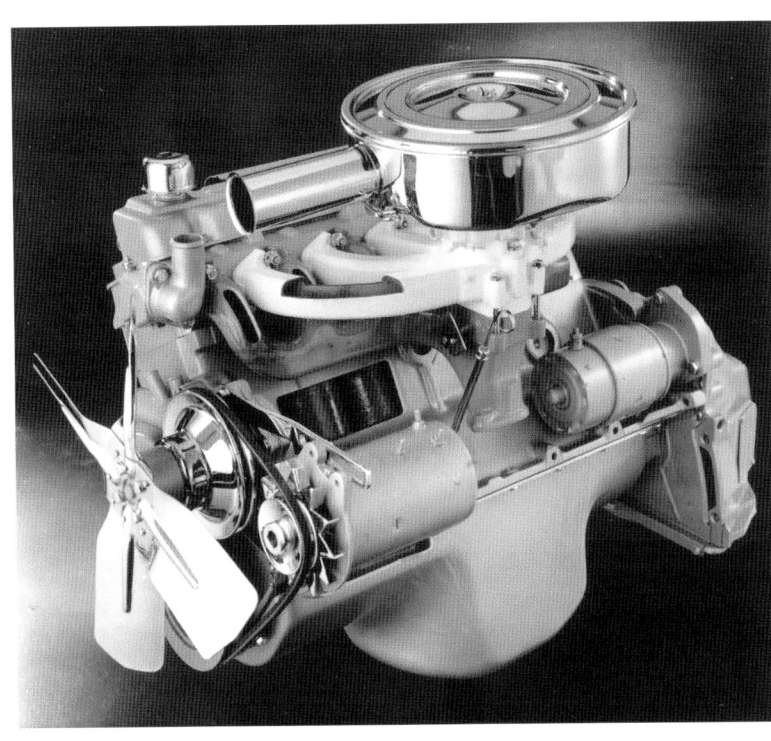

In der Basisvariante wurde der Dodge Dart 1960 mit einem 3,7-Liter-Sechszylin-der-Motor ausgeliefert

In the basic version, the Dodge Dart 1960 was shipped with a 6-cylinder 3.7-liter engine.

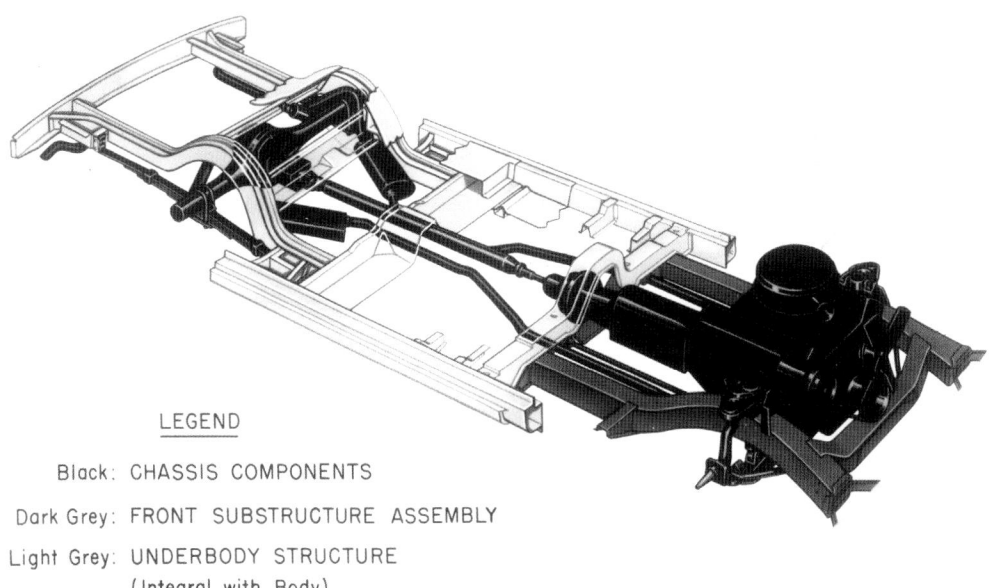

LEGEND

Black: CHASSIS COMPONENTS

Dark Grey: FRONT SUBSTRUCTURE ASSEMBLY

Light Grey: UNDERBODY STRUCTURE
(Integral with Body)

Schematische Darstellung des mehrteiligen Chassis-Aufbaus des Dodge Dart.
Diagram of the multipart chassis structure of the Dodge Dart.

Mit dem Dodge Dart debütierte 1960 ein auf Anhieb erfolgreiches Mittel-klassemodell des US-Herstellers. Als Basismotorisierung diente ein 3,7-Liter-Reihen-Sechszylinder.

The Dart introduced in 1960 was an immediately successful medium-class model of the US manufacturer Dodge. It was driven by a 3.7-liter 6-cylinder in-line engine in base.

VW Käfer 1200

Fahrzeugdaten

Hersteller:	Volkswagen
Land:	Deutschland
Modell:	Käfer 1200
Bauzeit:	1954–1964
Länge:	4.070 mm
Breite:	1.540 mm
Höhe:	1.500 mm
Radstand:	2.400 mm
Leergewicht:	730 kg
Antriebsart:	Hinterrad
Höchstgeschwindigkeit:	ca. 110 km/h
Verbrauch:	ca. 9 Liter/100 km

*Werte für Käfer 1200, 30 PS, Bj. 1960

Anfang der 1930er-Jahre erteilte der Reichsverband der Deutschen Automobilindustrie Ferdinand Porsche den Auftrag, einen preiswerten und sparsamen Pkw zu entwickeln. Er sollte 100 km/h schnell sein, nicht mehr als 7 Liter Treibstoff pro 100 Kilometer benötigen und weniger als 1.000 Reichsmark kosten. Bis zum Kriegsausbruch wurden vom offiziell als „Kraft-durch-Freude-Wagen" bezeichneten Käfer nur eine Handvoll Prototypen gebaut. Die Serienproduktion begann erst nach Kriegsende. Nach 58 Jahren und über 21,5 Millionen Einheiten lief 2003 der letzte Käfer vom Band. Damit ist er bis heute das weltweit zweitmeist gebaute Auto. Platz eins belegt der VW Golf. Der VW Käfer 1200 wurde 1953 vorgestellt. Der Hubraum von 1.131 cm³ war auf 1.192 cm³ erhöht worden, indem man die Zylinderbohrung von 75 Millimeter auf 77 Millimeter vergrößert hatte. Dadurch stieg die Leistung des Motors von 18 kW (25 PS) auf 22 kW (30 PS). Alleine bis 1957 wurden von dieser Version etwa 1,2 Millionen Stück verkauft.

Specifications

Manufacturer:	Volkswagen
Country:	Germany
Model:	Käfer (Beetle) 1200
Produced:	1954–1964
Length:	4070 mm
Width:	1540 mm
Height:	1500 mm
Wheelbase:	2400 mm
Empty weight:	730 kg
Type of drive:	Rear-wheel drive
Max. speed:	approx. 110 km/h
Fuel consumption:	approx. 9 l/100 km

* data for the Beetle 1200, 30 PS, built 1960

In the beginning of the 1930s, the Reich organization of German car manufacturers commissioned Ferdinand Porsche to develop a cheap and fuel-saving passenger car. It should reach a speed of 100 km/h, consume no more than 7 liters of fuel per 100 km and cost less than 1000 Reichsmark. Until the outbreak of WWII, only a handful of prototypes of "Käfer" (Beetle) were built – then officially called "Kraft durch Freude" car ("strength through joy," referring to the Nazi leisure organization). Series production only began after the war. After a production run of 58 years and more than 21.5 million units worldwide, the last Beetle left the assembly line in 2003. It is thus the second-most produced car behind the VW Golf. The VW Beetle 1200 was introduced in 1953. The displacement had been increased from 1131 cm³ to 1192 cm³ by extending the bore from 75 mm to 77 mm. This resulted in a power increase from 18 kW (25 hp) to 22 kW (30 hp). Until 1957 alone, 1.2 million units of this version were sold.

Der Motor

Motordaten

Bauart:	Viertakt-Boxermotor
Zylinderzahl:	4
Hubraum:	1.192 cm³
Bohrung:	77 mm
Hub:	64 mm
Leistung:	22 kW (30 PS)
	bei 3.400/min

Mit dem VW-Käfer erfuhr der Boxermotor seine größte Verbreitung. Seine Konstruktion veränderte sich während der gesamten Bauzeit kaum. Im Laufe der Jahrzehnte wurden im Wesentlichen nur der Hubraum, die Leistung und diverse Anbauteile verändert. Getreu dem historischen VW-Slogan „Luft kocht nicht, Luft gefriert nicht", war der Boxermotor des Käfers luftgekühlt. Ein über einen Keilriemen von der Kurbelwelle angetriebenes Lüfterrad sorgte für die benötigte Luftzirkulation um die Zylinder. Diese Art der Kühlung hatte einen entscheidenden Nachteil: Während sie bei hohen Drehzahlen für reichlich Frischluft sorgte, ließ sie bei geringen Drehzahlen zu wünschen übrig. Manchem Motor kostete deshalb nicht das Rasen auf der Autobahn, sondern das anschließende Stehen im Stau das Leben. Pro Zylinder waren je ein Ein- und Auslassventil eingebaut, die über eine unten liegende Nockenwelle sowie Stoßstangen und Kipphebel betätigt wurden.

The engine

Engine specifications

Type:	Four-stroke flat engine
Number of cylinders:	4
Displacement:	1192 cm³
Bore:	77 mm
Stroke:	64 mm
Power:	22 kW (30 hp)
	at 3400 rpm

The flat engine experienced its most extensive use with the VW Beetle. During the whole production period of the Beetle, the engine design remained nearly unchanged. Over the course of the decades, only the displacement, the power and various supplemental parts were altered. According to the historic VW slogan, "Air does neither boil nor freeze!" the Beetle had an air-cooled flat engine. A fan wheel driven by the crankshaft via a V-belt provided the required air circulation around the cylinders. However, this type of cooling had an essential disadvantage. While it provided enough fresh air at high engine speeds, it left a lot to be desired at low engine speed. It was not so much careening on the autobahn that killed one or the other engine, but the ensuing traffic jam. Each cylinder had one intake and outlet valve, which were actuated by a bottom-mounted camshaft, push rods and rocker levers.

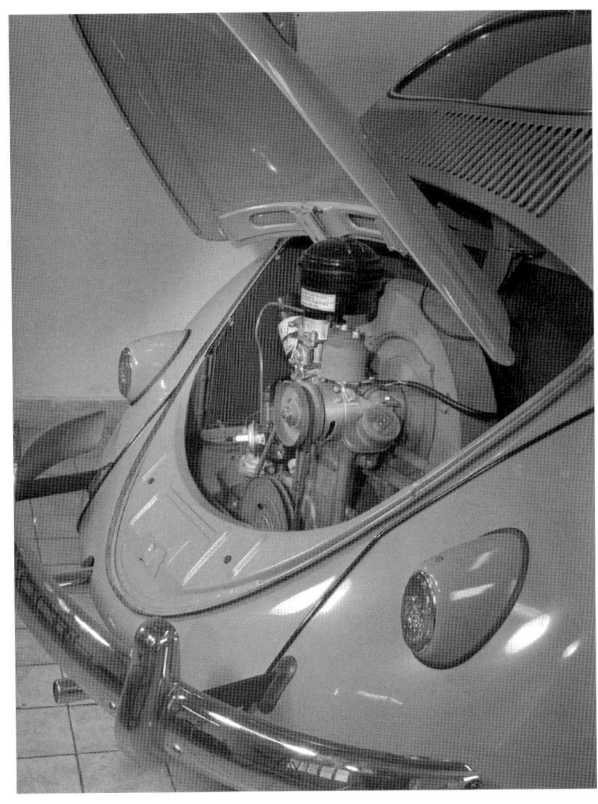

Der VW Käfer wurde von einem Boxermotor angetrieben, der in seiner Grundkonstruktion unverändert blieb.
The VW Beetle was driven by a flat engine which remained basically unmodified.

Der VW Käfer ist zweifelsohne weit mehr als nur ein Auto. Er ist ein Stück deutscher Wirtschaftsgeschichte, das ganze Generationen von Autofahrern nachhaltig geprägt hat.

The VW Beetle is without doubt more than just a car. It is a piece of German economic history and has put its mark on generations of drivers.

Der Renault 4 setzte mit vier Türen, großer Heckklappe und variablem Innenraum Maßstäbe in seiner Klasse.

With its four doors, large hatch and variable passenger compartment, the Renault 4 set new standards in its class.

Renault 4

Fahrzeugdaten

Hersteller:	Renault
Land:	Frankreich
Modell:	4
Bauzeit:	1961–1992
Länge:	3.661 mm
Breite:	1.485 mm
Höhe:	1.550 mm
Radstand:	2.400–2.450 mm
Leergewicht:	600 kg
Antriebsart:	Vorderrad
Höchstgeschwindigkeit:	110 km/h
Verbrauch:	5,9 Liter/100 km

*Werte für 1961 Renault 4

Der französische Kleinwagen setzte in seinem Marktsegment ab 1961 neue Maßstäbe: Der Renault 4 war die weltweit erste Kombi-Limousine mit vier Türen, großer Heckklappe und variablem Innenraum. Der erste Renault-Personenwagen mit Frontantrieb wurde zu einem Erfolgsmodell, denn bis 1992 liefen mehr als acht Millionen Exemplare vom Band. Allein in Deutschland verkaufte Renault mehr als 900.000 Fahrzeuge vom Typ R4.

Die Gründe für die Zuneigung der Käuferschaft waren einfach: Der Renault 4 war praktisch, zuverlässig und preiswert. Der 3.661 Millimeter lange Wagen bot Platz für fünf Erwachsene plus Reisegepäck. Das Kofferraumvolumen betrug mindestens 255 Liter, ließ sich aber auf beeindruckende 950 Liter erweitern. Denn der R4 war das erste Serienauto, bei dem sich die Rückbank komplett zusammenfalten und nach vorne klappen ließ. So entstand ein gut nutzbarer Laderaum. Und mit einer Höhe von 155 Zentimeter hatte der Renault fast die Maße eines heutigen Kompakt-Vans.

Specifications

Manufacturer:	Renault
Country:	France
Model:	4
Produced:	1961–1992
Length:	3661 mm
Width:	1485 mm
Height:	1550 mm
Wheelbase:	2400–2450 mm
Empty weight:	600 kg
Type of drive:	Front-wheel drive
Max. speed:	110 km/h
Fuel consumption:	5.9 l/100 km

* data for the 1961 Renault 4

From 1961 on, this French subcompact car set new standards in its market segment. The Renault 4 was the world's first station wagon with four doors, a large hatch and variable passenger compartment. The first front-driven Renault passenger car was a huge success. Until 1992, more than 8 million vehicles left the assembly line. In Germany alone Renault sold more than 900,000 R4.

The customers liked the car for simple reasons: The Renault 4 was practical, reliable and inexpensive. At a length of 3661 mm, the car accommodated five adults plus luggage. The trunk had a volume of at least 255 liter and could be extended to an impressive 950 liters, as the R4 was the first production car where the rear bench could be completely folded away to create a large luggage compartment. With its height of 155 cm, the Renault nearly had the dimensions of a compact van.

Der Motor

Motordaten

Bauart:	Viertakt-Reihenmotor
Zylinderzahl:	4
Hubraum:	747 cm³
Bohrung:	54,5 mm
Hub:	80 mm
Leistung:	17 kW (24 PS)
	bei 4.500/min

Im ersten Renault 4, der auf der Internationalen Automobil-Ausstellung (IAA) in Frankfurt am Main 1961 vorgestellt wurde, war ein unauffälliger Vierzylinder-Reihenmotor mit 0,75 Litern Hubraum verbaut. Seine 17 kW (24 PS) genügten, um den 600 Kilogramm leichten Wagen auf bis zu 105 km/h zu beschleunigen. Die Maximalleistung gab der Motor bei 4.500 Umdrehungen pro Minute ab. Sein höchstes Drehmoment von 55 Nm erreichte er bei 2.000/min. Der Vierzylinder besaß eine seitliche Nockenwelle, pro Zylinder verfügte er über zwei hängende Ventile. Das Verdichtungsverhältnis lag bei 8,5:1. Der vorne eingebaute Motor wirkte über ein Dreigang-Schaltgetriebe auf die Vorderräder. In späteren Modelljahren bot Renault auch stärkere Triebwerke mit bis zu 25 kW (34 PS) an. Mit dieser Top-Motorisierung erreichte der Renault 4 eher gemächliche 120 km/h. Bescheiden gab sich der Kleinwagen bis zu seinem Produktionsende 1992 auch beim Verbrauch, denn je nach Motorisierung benötigte er zwischen 5,4 und 6,5 Liter pro 100 Kilometer.

The engine

Engine specifications

Type:	Four-stroke in-line engine
Number of cylinders:	4
Displacement:	747 cm³
Bore:	54.5 mm
Stroke:	80 mm
Power:	17 kW (24 hp)
	at 4500 rpm

The first Renault 4, which was introduced at the International Motor Show (IAA) in Frankfurt/Main in 1961, had an inconspicuous 4-cylinder in-line engine with a displacement of 0.75 liters. Its 17 kW (24 hp) were sufficient to accelerate the 600 kg car up to 105 km/h. The maximum power was reached at 4500 rpm, the maximum torque of 55 Nm at 2000 rpm. The engine had a side-mounted camshaft and two overhead valves per cylinder. The compression ratio amounted to 8.5:1. The power of the front-installed engine was transferred to the front wheels via a manual three-gear transmission. In later model years, Renault also offered more powerful engines with up to 25 kW (34 hp), which could accelerate the Renault 4 to a rather leisurely pace of 120 km/h. Up to end of production in 1992, the small car was also very humble when it came to fuel consumption. Depending on the selected engine, it needed between 5.4 and 6.5 liters per 100 km.

Die ersten Renault 4 besaßen einen Vierzylinder-Reihenmotor mit 0,75 Litern Hubraum.
The first Renault 4 had a 4-cylinder in-line engine with a displacement of 0.75 liters.

Der 1961 vorgestellte Renault 4 entwickelte sich zu einem großen Verkaufserfolg für den französischen Hersteller. Bis zum Produktionsende 1992 liefen
mehr als acht Millionen Exemplare des R4 vom Band.

The Renault 4, introduced in 1961, became a bestseller for the French manufacturer. Until the end of production in 1992, more than 8 million vehicles
left the assembly line.

„Neue Klasse": Die Limousine verband Komfort mit Fahrdynamik – und gilt als Vorläufer der 5er-Reihe.

"New class:" The sedan combined comfort and a dynamic ride and is thus seen as a predecessor of the series 5.

BMW 1500

Fahrzeugdaten

Hersteller:	BMW
Land:	Deutschland
Modell:	1500
Bauzeit:	1961–1964
Länge:	4.427 mm
Breite:	1.651 mm
Höhe:	1.427 mm
Radstand:	2.550 mm
Leergewicht:	1.060 kg
Antriebsart:	Hinterrad
Höchstgeschwindigkeit:	150 km/h
Verbrauch:	10 Liter/100 km

„Neue Klasse" – mit diesem Begriff verband sich der BMW 1500, der 1961 auf der Internationalen Automobil Ausstellung in Frankfurt am Main präsentiert wurde. Die neue Limousine mit der internen Bezeichnung „E115" wurde bei ihrer Premiere von Fachpresse und Publikum begeistert aufgenommen. BMW zielte mit dem 1500 auf jene Marktlücke sportlich-dynamischer Mittelklasse-Modelle, die die inzwischen zahlungsunfähige Firma Borgward mit der Isabella besetzt hatte. Erst im Oktober 1962 begann die Serienproduktion des BMW. Zum Verkaufsstart wurde die viertürige Limousine für 9.485 DM angeboten. Anfangs litt die Neukonstruktion unter zahlreichen Kinderkrankheiten. Trotzdem konnten die Münchner bis Ende 1964 fast 24.000 Exemplare verkaufen, bevor das Modell durch den BMW 1600 abgelöst wurde. Der Erfolg der „Neuen Klasse" sicherte die Zukunft von BMW als eigenständige Marke, nachdem wenige Jahre zuvor ein Verkauf an Daimler-Benz nur knapp verhindert worden war.

Specifications

Manufacturer:	BMW
Country:	Germany
Model:	1500
Produced:	1961–1964
Length:	4427 mm
Width:	1651 mm
Height:	1427 mm
Wheelbase:	2550 mm
Empty weight:	1060 kg
Type of drive:	Rear-wheel drive
Max. speed:	150 km/h
Fuel consumption:	10 l/100 km

"New class" was the term associated with the BMW 1500, which was introduced at the IAA in Frankfurt/Main in 1961. At its debut, the new sedan with the internal designation "E115" was received enthusiastically by the trade press and the audience. With the 1500, BWM targeted the market gap of sporty, dynamic middle class models, which was previously served by the now insolvent Borgward company with its Isabella model. Series production of the new BMW did start not earlier than in October 1962. At market launch, the four-door sedan was quoted at 9485 DM. In the beginning, the new design suffered from several teething troubles. Nonetheless the company could sell nearly 24.000 units until the end of 1964, when the model was replaced by the BMW 1600. The success of the "new class" secured the future of BMW as an independent brand. A few years ago, a sale to Daimler-Benz had only narrowly been avoided.

Der Motor

Motordaten

Bauart:	Viertakt-Reihenmotor
Zylinderzahl:	4
Hubraum:	1.499 cm³
Bohrung:	82 mm
Hub:	71 mm
Leistung:	59 kW (80 PS) bei 5.700/min

Der BMW 1500 war mit einem neu entwickelten Vierzylinder-Reihenmotor bestückt. Der Motorblock bestand aus Grauguss, während der Zylinderkopf aus Aluminium gefertigt war. Der Motor verfügte über eine oben liegende Nockenwelle sowie zwei Ventile pro Zylinder. Der Hubraum lag bei 1,5 Litern. Bei einer Drehzahl von 5.700 Umdrehungen pro Minute leistete das BMW-Aggregat 59 kW (80 PS). Die Konstruktion mit der internen Bezeichnung „M10" erwies sich als ausgesprochen erfolgreich und wurde erst Ende der 1980er durch eine neue BMW-Motorengeneration abgelöst. Auch im Motorsport zeigte sich die Leistungsfähigkeit dieses Motors: Er lieferte die Basis für verschiedene Rennmotoren. Unter anderem basierte der Turbomotor, mit dem BMW ab 1982 als Motorenlieferant in der Formel 1 antrat, auf dem Motorblock des 1961 vorgestellten 1500. 1983 gewann Nelson Piquet mit einem Brabham-BMW die Formel-1-Weltmeisterschaft. Der BMW-Turbomotor leistete bis zu 1.027 kW (1.400 PS).

The engine

Engine specifications

Type:	Four-stroke in-line engine
Number of cylinders:	4
Displacement:	1499 cm³
Bore:	82 mm
Stroke:	71 mm
Power:	59 kW (80 hp) at 5700/min

The BMW 1500 was equipped with a newly developed 4-cylinder in-line engine. The engine block consisted of grey cast iron, while the cylinder head was made of aluminum. The engine had an overhead camshaft, two valves per cylinder and a displacement of 1.5 liters and provided a power of 59 kW (80 PS) at 5700 rpm. The design with the internal designation "M10" proved to be extraordinarily successful and was replaced by a new generation of BMW engines not before the end of the 1980s. The performance of the engine also became obvious in motorsports, as it served as the basis for various racing engines. Among others, the turbo engine that BMW used in Formula 1 racing since 1982 was based on the engine block of the 1500 introduced in 1961. In 1983, Nelson Piquet won the Formula 1 world championship aboard a Brabham BMW. Its turbo engine provided up to 1027 kW (1400 hp).

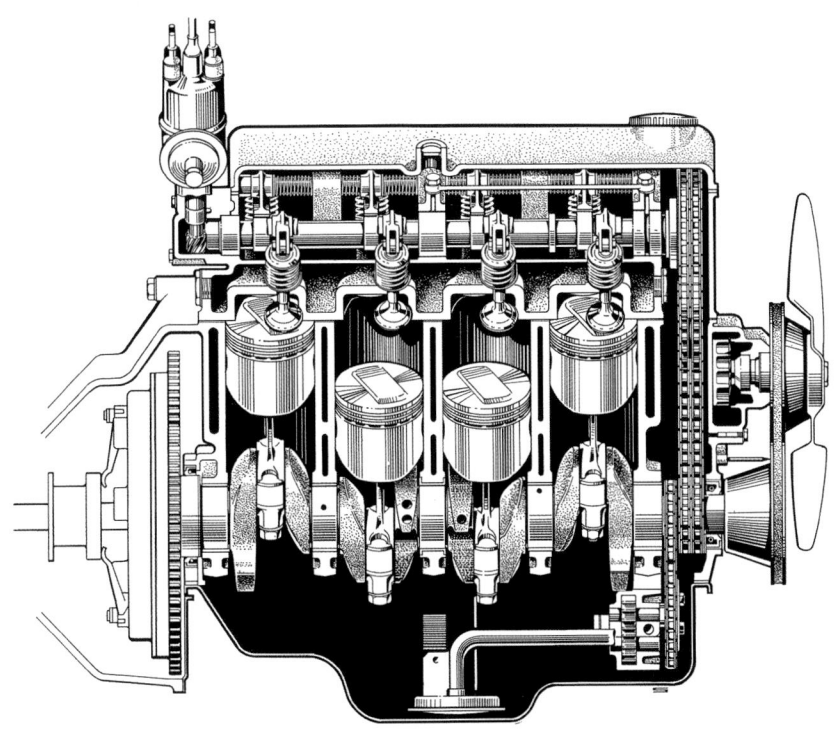

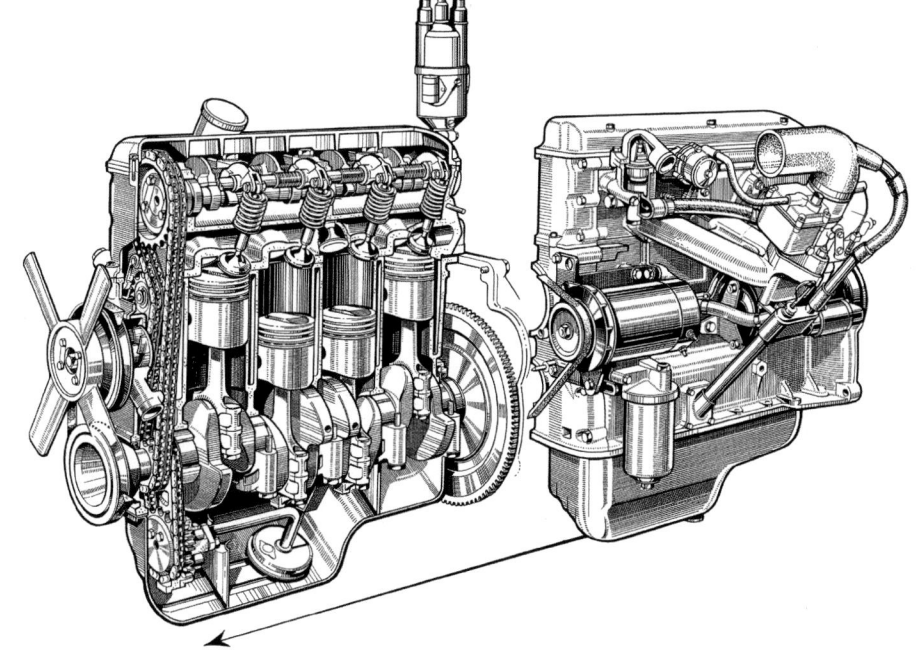

Der Reihen-Vierzylinder mit der internen Bezeichnung „M10" kam erstmals im BMW 1500 zum Einsatz.

The 4-cylinder in-line engine with the internal designation "M10" was first used in the BMW 1500.

Der BMW-Motor verfügte über eine oben liegende Nockenwelle und zwei Ventile pro Zylinder.

The BMW engine had an overhead camshaft and two valves per cylinder.

Das Herzstück des BMW 1500 war 1961 der neue Reihen-Vierzylinder. Dieser Motor zeigte sich sehr entwicklungsfähig und lieferte auch die Basis für das Formel-1-Turbo-Triebwerk der 1980er-Jahre.

The heart of the BMW 1500 was the new 4-cylinder in-line engine. This engine proved to be highly capable of further development and also served as the basis for the Formula 1 turbo engine of the 1980s.

1963 präsentierte Mercedes-Benz den neuen Typ 600. Das Modell verfügte über einen 6,3-Liter-V8-Motor.
In 1963, Mercedes-Benz introduced the new type 600 with a 6.3-liter V8 engine.

Mercedes-Benz 600

Fahrzeugdaten

Hersteller:	Mercedes-Benz
Land:	Deutschland
Modell:	600
Bauzeit:	1963–1981
Länge:	5.540 mm
Breite:	1.950 mm
Höhe:	1.500 mm
Radstand:	3.200 mm
Leergewicht:	2.475 kg
Antriebsart:	Hinterrad
Höchstgeschwindigkeit:	207 km/h
Verbrauch:	24 Liter/100 km

Der neue Mercedes-Benz 600 war bei seiner Premiere 1963 die Sensation auf der IAA in Frankfurt am Main: Die Marke mit dem Stern hatte eine repräsentative Limousine entwickelt, die Komfort und Luxus mit den Fahrleistungen eines Sportwagens kombinierte. Mit dem 600 zeigte Mercedes-Benz dem staunenden Publikum den Stand der technischen Möglichkeiten: V8-Motor mit 184 kW (250 PS), Luftfederung, ein hydraulisches Servosystem, eine Klimaanlage und hydraulische Fensterheber.

Und so war es kein Wunder, dass die Prominenz jener Zeit gerne im Mercedes-Benz 600 Platz nahm – die Liste reichte von Ex-Beatle John Lennon über Sowjetführer Leonid Breschnew bis zu Papst Paul VI. Damit erfüllte dieses Modell erfolgreich seine Aufgabe als Imageträger für Mercedes-Benz. Wirtschaftlich soll der Mercedes-Benz 600 dagegen bis zum Ende der Produktion im Jahr 1981 ein Verlustbringer für das Stuttgarter Unternehmen geblieben sein. Insgesamt entstanden lediglich 2.677 Exemplare.

Specifications

Manufacturer:	Mercedes-Benz
Country:	Germany
Model:	600
Produced:	1963–1981
Length:	5540 mm
Width:	1950 mm
Height:	1500 mm
Wheelbase:	3200 mm
Empty weight:	2475 kg
Type of drive:	Rear-wheel drive
Max. speed:	207 km/h
Fuel consumption:	24 l/100 km

At its debut in 1963, the new Mercedes-Benz 600 caused a sensation at the IAA in Frankfurt/Main. The "star brand" had developed a limousine that combined comfort and luxury with the driving characteristics of a sports car. With the 600, Mercedes-Benz demonstrated to a marveling audience what state-of-the-art technology could be achieved: a V engine with 184 kW (250 hp), air cushioning, hydraulic servo system, air conditioning and hydraulic windows lifters.

Thus it was no wonder that contemporary celebrities loved to climb aboard a Mercedes-Benz 600, from ex-Beatle John Lennon to Soviet leader Leonid Brezhnev to Pope Paul VI. Thus the model successfully fulfilled its role as brand ambassador. Economically, however, the Mercedes-Benz 600 proved to be a loss-making business up to the end of production in 1981. In total, only 2677 units were built.

Der Motor

Motordaten

Bauart:	Viertakt-90-Grad-V-Motor
Zylinderzahl:	8
Hubraum:	6.332 cm³
Bohrung:	103 mm
Hub:	95 mm
Leistung:	184 kW (250 PS)
	bei 4.000/min

Der 6,3-Liter-V8-Motor mit der Bezeichnung „M100" machte den Mercedes 600 bei seiner Markteinführung zu einer der schnellsten Serienlimousinen der Welt. Immerhin erreichte das Antriebsaggregat des „Großen Mercedes" 250 PS bei 4.000/min. Die beiden Zylinderbänke des Achtzylinder-Triebwerks standen im Winkel von 90 Grad zueinander. Der Kurbelgehäuse bestand aus Grauguss, während die Zylinderköpfe aus Aluminium gefertigt waren. Die Brennräume der Zylinder waren keilförmig ausgeführt – pro Zylinder besaß der Motor je zwei Ventile. In seiner ursprünglichen Ausführung war der M100 mit einer mechanischen Benzineinspritzung versehen. Der leistungsstarke Motor kam nicht nur im Mercedes-Benz 600 zum Einsatz, denn zwischen 1968 und 1972 trieb er auch den Typ 300 SEL 6,3 an. Ab 1975 sorgte der M100 auch im 450 SEL 6,9 für Vortrieb. Der Hubraum des Top-Modells der damaligen S-Klasse war auf 6,9 Liter angewachsen, die Leistung betrug nun 210 kW (286 PS).

The engine

Engine specifications

Type:	Four-stroke 90° V engine
Number of cylinders:	8
Displacement:	6332 cm³
Bore:	103 mm
Stroke:	95 mm
Power:	184 kW (250 hp)
	at 4000 rpm

The 6.3-liter V8 engine designated as "M100" made the Mercedes 600 to one of the world's fastest series at it launch to market. It delivered as much as 250 hp at 4000 rpm. The two cylinder banks were arranged at an angle of 90°. The crankcase was made of grey cast iron while the cylinder heads were made of aluminum. The combustion chambers of the cylinders were wedge-shaped. Each cylinder had two valves. The original version of the M100 used mechanical fuel injection. This powerful engine was not only used in the Mercedes-Benz 600 but also drove the 300 SEL 6.3 between 1968 and 1972 and the 450 SEL 6.9 beginning with 1975. The displacement of the then top model of the S class had been increased to 6.9 liters, delivering a power of 210 kW (286 hp).

Der 6,3-Liter-V8-Motor des Mercedes-Benz 600 leistete 184 kW (250 PS) bei 4.000/min.

The 6.3-liter V8 engine of the Mercedes-Benz 600 provided 184 kW (250 hp) at 4000 rpm.

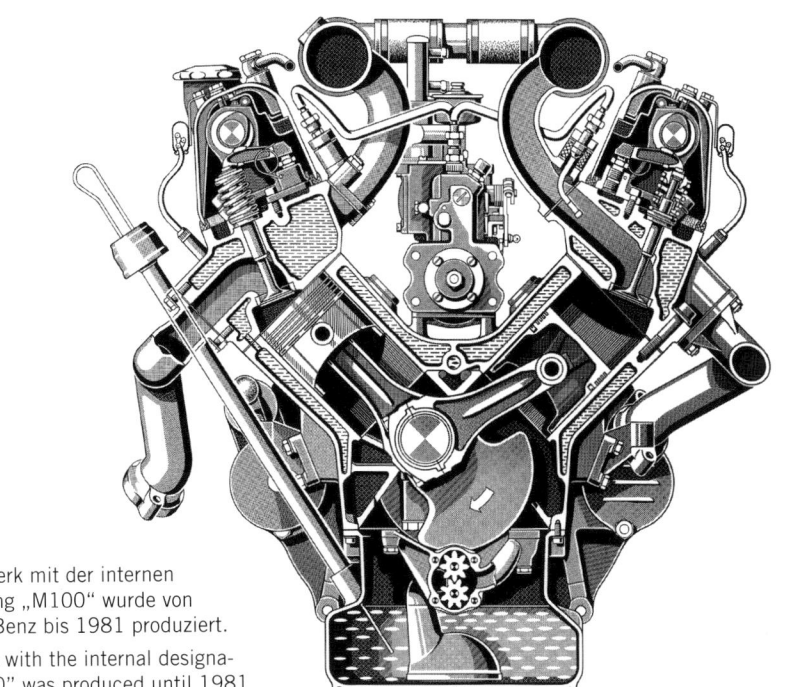

Das Triebwerk mit der internen Bezeichnung „M100" wurde von Mercedes-Benz bis 1981 produziert.

The engine with the internal designation "M100" was produced until 1981.

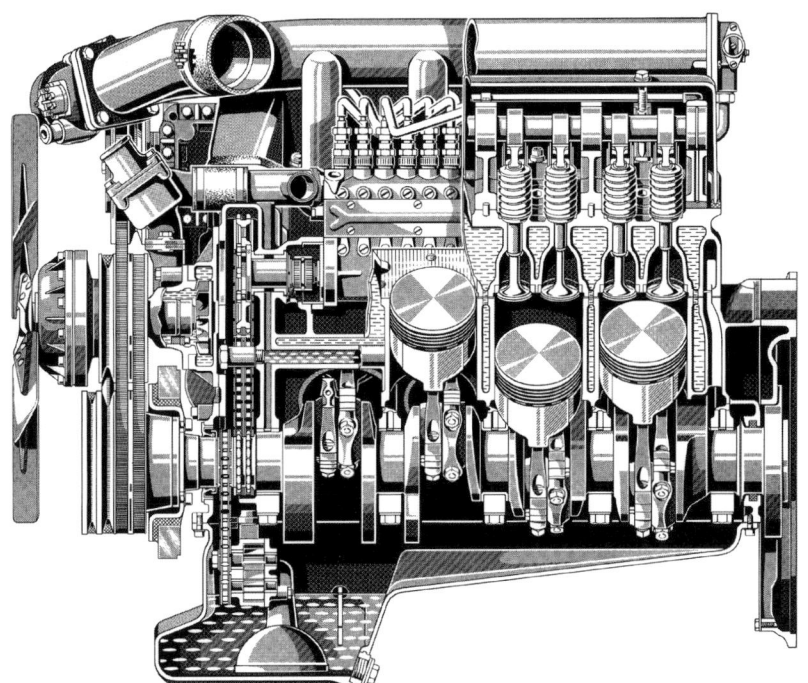

Mit dem Typ 600 meldete sich Mercedes-Benz in der Liga der Luxus-Hersteller zurück. Die Limousine besaß einen V8-Motor, war 200 km/h schnell – und bot alle Annehmlichkeiten, die 1963 technisch machbar waren.

With the type 600, Mercedes-Benz made its comeback in the league of the manufacturers of luxury cars. The sedan had a V8 engine, a top speed of 200 km/h and provided all amenities that were technologically feasible in 1963.

Mit dem Porsche 911 2.0 Coupé brach 1963
eine neue Zeitrechnung im Sportwagenbau an.

With the Porsche 911 2.0 coupé, a new era of
sports car design began in 1963.

Porsche 911

Fahrzeugdaten

Hersteller:	Porsche
Land:	Deutschland
Modell:	911
Bauzeit:	1963–1973
Länge:	4.163 mm
Breite:	1.610 mm
Höhe:	1.321 mm
Radstand:	2.210 mm
Leergewicht:	1.080 kg
Antriebsart:	Hinterrad
Höchstgeschwindigkeit:	211 km/h
Verbrauch:	13,5 Liter/100 km

*Werte für 911, Bj. 1964

Mit dem 911 schuf Porsche 1963 die Ikone unter den Sportwagen schlechthin. Die Modellbezeichnung „911" ergab sich auf Umwegen. Zunächst hörte der neue Sportwagen aus Zuffenhausen auf den Namen „901". Dies war jedoch eine Ziffernfolge, deren Kombination mit der Null in der Mitte Peugeot für sich beanspruchte. Also änderte Porsche die Bezeichnung kurzerhand in „911". Verglichen mit dem Porsche 356 setzte zwar auch der neue 911 auf das bewährte Prinzip eines luftgekühlten Heckmotors, doch handelte es sich in sämtlichen Belangen um ein völlig neues Fahrzeug. Bei der Entwicklung waren eine modernere Fahrwerkstechnik, ein vergrößertes Raumangebot und eine drastisch optimierte Fahrdynamik primäre Zielsetzungen. Gleichzeitig sollte der neue 911 in seiner Formensprache aber auf den ersten Blick als Porsche erkennbar sein, was Designer Ferdinand Alexander Porsche perfekt gelang. Bis heute in der siebten 911-Generation hat sich an den grundsätzlichen Grundzügen der Karosserie nichts geändert.

Specifications

Manufacturer:	Porsche
Country:	Germany
Model:	911
Produced:	1963–1973
Length:	4163 mm
Width:	1610 mm
Height:	1321 mm
Wheelbase:	2210 mm
Empty weight:	1080 kg
Type of drive:	Rear-wheel drive
Max. speed:	211 km/h
Fuel consumption:	13.5 l/100 km

* data for the 911 built in 1964

With the 911 introduced in 1963, Porsche created the epitome of a sports car. The model designation "911," however, was arrived at only via a detour. First, the new sports car from Zuffenhausen was called "901," but Peugeot claimed the right to use three-digit numbers with a middle zero. Thus, the number was changed to "911" without further ado. Similar to the 356, the 911 also relied on the concept of an air-cooled rear-mounted engine, but in all other aspects, it was a completely different vehicle. The main objectives for development had been state-of-the-art undercarriage technology, more space and dramatically improved driving characteristics, but at the same time, the shape of the new 911 should be easily recognizable as a Porsche. The latter objective was perfectly achieved by designer Ferdinand Alexander Porsche. Today, the seventh generation still uses basically the same auto body design.

Der Motor

Motordaten

Bauart:	Viertakt-Boxermotor
Zylinderzahl:	6
Hubraum:	1.991 cm³
Bohrung:	80 mm
Hub:	66 mm
Leistung:	96 kW (130 PS)
	bei 6.100/min

Herzstück des neuen Porsche 911 war der Sechszylinder-Boxermotor, der in guter Tradition des Hauses gebläsegekühlt war. Aus 2 Litern Hubraum schöpfte das Triebwerk 96 kW (130 PS), die bei 6.100 Touren abgegeben wurden. Bereits bei 4.200 Umdrehungen pro Minute entfaltete der Sechszylinder sein maximales Drehmoment von 174 Nm. Während der Vierzylinder-Boxer des Porsche 356 noch über eine OHV-Steuerung und damit eine unten liegende Nockenwelle, Stoßstangen und Kipphebel verfügte, setzte Konstrukteur Hans Mezger beim „Elfer"-Motor auf modernere, drehzahlfestere Technik. So verfügte der Boxer des 911 über je eine oben liegende Nockenwelle pro Zylinderreihe. Die Betätigung der beiden Ventile pro Brennraum übernahmen Kipphebel. „Eine durch und durch solide Konstruktion", gab Hans Mezger zu Protokoll, und tatsächlich ging das in seinen Grundzügen kaum veränderte Triebwerk erst mit der letzten luftgekühlten 911-Baureihe vom Typ 993 Ende der 1990er-Jahre in Rente.

The engine

Engine specifications

Type:	Four-stroke flat engine
Number of cylinders:	6
Displacement:	1991 cm³
Bore:	80 mm
Stroke:	66 mm
Power:	96 kW (130 hp)
	at 6100 rpm

The heart of the new Porsche 911 was a 6-cylinder flat engine with the traditional air cooling of the manufacturer. With a displacement of 2 liters, the engine provided 96 kW (130 hp) at 6100 rpm. The maximum torque of 174 Nm was already achieved at 4200 rpm. While the 4-cylinder flat engine of the Porsche 356 still had an OHV control with a bottom mounted camshaft, push rods and rocker lever, constructing engineer Hans Mezger used a more modern technology with a more stable engine speed for the new model. Thus the flat engine of the 911 was equipped with one overhead camshaft per cylinder bank. The two valves per combustion chamber were actuated by rocker levers. "A thoroughly robust construction," Hans Mezger noted. The basic design of the engine was virtually unaltered when it went into retirement with the last air-cooled 911 series type 993 at the end of the 1990s.

Der 2-Liter-Boxer-Motor des 911. In dieser frühen Version besaß das Triebwerk Solex-Vergaser.

The 2.0-liter flat engine of the 911. In this early version, the engine had Solex carburetors.

Der 2-Liter-Boxermotor des Porsche 911 2.0 Coupé von 1964 leistete muntere 130 PS.

The 2-liter flat engine of the 1964 Porsche 911 2.0 Coupé provided vivid 130 hp.

Seit 1963 ist die Tradition des Porsche 911 mit Heckmotor und der typischen Karosserieform ungebrochen. Einen Umbruch könnte jedoch der GT-Rennwagen 911 RSR von 2017 mit Mittelmotor darstellen.

Since 1963, the tradition of the Porsche 911 with rear-mounted engine and the typical, unique shape is unabated. However, the GT racing car 911 RSR introduced in 2017 with a midship engine could herald a change.

Ford Mustang Boss 302 aus dem Jahr 1969. Sein 5-Liter-V8-Motor leistete 216 kW (290 PS).
Ford Mustang Boss 302 of 1969. The 5-liter V8 engine provided 216 kW (290 hp).

Der wunderschön gezeichnete Fastback zählt zu den begehrtesten Mustang-Modellen überhaupt.
The beautifully shaped Fastback is one of the most sought-after Mustang models of all times.

Ford Mustang I

Fahrzeugdaten

Hersteller:	Ford
Land:	USA
Modell:	Mustang I
Bauzeit:	1964–1973
Länge:	4.760 mm
Breite:	1.824 mm
Höhe:	1.250 mm
Radstand:	2.700 mm
Leergewicht:	1.462 kg
Antriebsart:	Hinterrad
Höchstgeschwindigkeit:	195 km/h
Verbrauch:	15–22 Liter/100 km

*Werte für Mustang Coupé 1969

Kaum ein US-Car wurde im Rahmen der Markteinführung so ins Rampenlicht gestellt wie der Ford Mustang, als er am 17. April 1964 auf der New Yorker Weltausstellung seinen Verkaufsstart feierte. Für Ford war der Mustang von eminenter Wichtigkeit, versprach man sich doch hohe Verkaufszahlen. Nach dem Motto „kompakt, aber kraftvoll motorisiert" hatte Ford mit dem Mustang – angelehnt an das galoppierende Pferd als Markensymbol – die neue Fahrzeugklasse der „Pony Cars" geschaffen. Mit dem Debüt des Mustang als Coupé sowie als Cabriolet im Jahr 1964 einher gingen zwei Motorvarianten. So waren ein Reihensechszylinder sowie ein typisch amerikanischer V8 verfügbar. Ein Jahr später, 1965, folgte mit der Fastback-Variante eine sportiv gezeichnete Fließheck-Version. Eine Besonderheit stellten die vor September 1964 gebauten Mustang-Modelle dar, die in Szenekreisen kurz als „Modelljahr 1964 ½" bezeichnet werden. Optisch sind sie am kleineren Kühlergrill zu erkennen.

Specifications

Manufacturer:	Ford
Country:	USA
Model:	Mustang I
Produced:	1964–1973
Length:	4760 mm
Width:	1824 mm
Height:	1250 mm
Wheelbase:	2700 mm
Empty weight:	1462 kg
Type of drive:	Rear-wheel drive
Max. speed:	195 km/h
Fuel consumption:	15–22 l/100 km

* data for the Mustang coupé 1969

Nearly no other US car enjoyed the limelight during market launch as much as the Ford Mustang at the New York World's Fair on April 17th, 1964. The Mustang was of eminent importance for Ford, as the company expected high sales figures. Based on the motto "compact but powerful" and inspired by the galloping horse of the brand logo, Ford had created the new vehicle class of "pony cars" starting with the Mustang. At the debut of the Mustang as a coupé and a convertible in 1964, two engine types were available, namely a 6-cylinder in-line engine and a typical American V8. In the following year 1965, the sporty fastback model followed. The Mustang models built before September 1964 are a peculiarity. Among the fans, they are simply called "model year 1964 ½." They can be recognized by their smaller radiator grille.

Der Motor

Motordaten

Bauart:	Viertakt-90-Grad-V-Motor
Zylinderzahl:	8
Hubraum:	4.949 cm³
Bohrung:	102 mm
Hub:	76 mm
Leistung:	216 kW (290 PS)
	bei 5.200/min

Für einen gehörigen Schuss Rennsport-Feeling sorgte der Ford Mustang Boss 302 in der dritten von insgesamt vier Generationen des Mustang, Serie I aus den Jahren 1969 bis 1970. In ihm arbeitete ein von der Rennsportversion der Trans-Am-Serie abgeleiteter V8-Motor mit 5 Litern Hubraum. Ein noch sehr viel wuchtigerer Motor wurde mit dem Boss 429 mit 7.031 cm³ Hubraum angeboten, der allerdings über dem Hubraum-Limit der Trans-Am-Serie von 5 Litern lag. Die maximale Leistung von 216 kW (290 PS) gab der 302-Cubic-Inch-Motor bei 5.200 Umdrehungen pro Minute ab. Der Boss 429 zeigte sich mit 7 Litern Hubraum jedoch noch wesentlich imposanter. Mit seinen 287 kW (390 PS) bei 5.200 Umdrehungen pro Minute bot er nicht nur sehr viel mehr Leistung, sondern lieferte mit 665 Nm vor allem auch ein beeindruckend souveränes Drehmoment. Sowohl beim Boss 302 wie auch beim 429 wurde das Drehmoment über ein Viergang-Schaltgetriebe an die Hinterräder weitergegeben.

The engine

Engine specifications

Type:	4-stroke 90° V-engine
Number of cylinders:	8
Displacement:	4949 cm³
Bore:	102 mm
Stroke:	76 mm
Engine power:	216 kW (290 hp)
	at 5200 rpm

The Ford Mustang Boss 302 provided for a proper amount of racing feeling. It belonged to the third of four generations of the Mustang Series I, built in the years 1969 to 1970. It used a V8 engine with a displacement of 5 liters, developed from the racing version of the Trans-Am series. The Boss 429 had an even more powerful engine with a displacement of 7031 cm³; however, it thus exceeded the 5-liter displacement limit of the Trans-Am races. At 5200 rpm, this 302 cubic inch engine provided a maximum power of 216 kW (290 hp). The Boss 429 with a displacement of 7 liters was even more impressive. With 287 kW (390 hp) at 5200 rpm, it was not only more powerful but also provided an impressive torque of 665 Nm. The Boss 302 as well as the 429 used a four-gear manual transmission to transfer the torque to the rear wheels.

Der 216 kW (290 PS) starke Boss 302 wurde auf Basis eines Rennsportmotors entwickelt.

The Boss 302 with 216 kW (290 hp) was based on a racing engine.

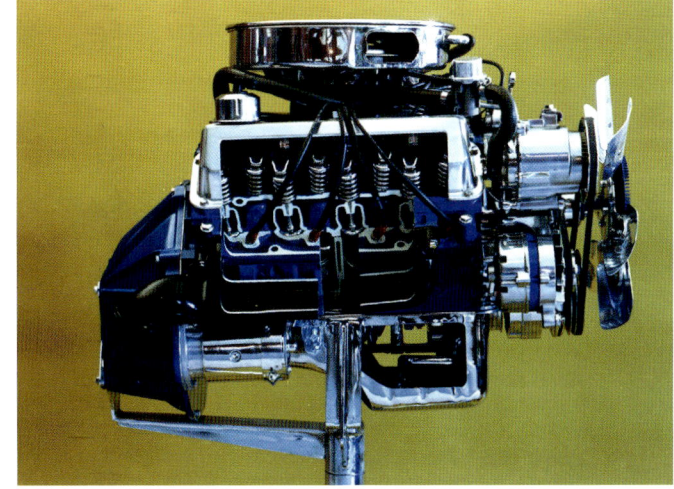

Der Ford T5 betont mehr denn je die sportliche Note. Das ist das praktische Resultat der langen Liste erfolgreicher Rallye-Fahrten. Oder zieht es Sie mehr zu der ganz eigenen Welt der Thunderbirds? Das automatische Cruise-O-Matic-Getriebe dieses Wagens bringt die enorme Kraft eines 255 DIN-PS starken V8-Motors weich auf die Straße. Im Thunderbird Convertible verschwindet das Dach vollkommen unter dem glattflächigen Heck.

FORD T5

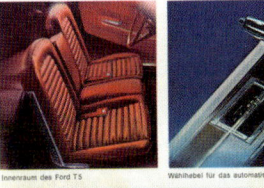

Innenraum des Ford T5

Wählhebel für das automatische Getriebe

Ford T5 Convertible, 2türig

Ford T5 Fastback 2+2

FORD THUNDERBIRD

Die mächtigen Zwillingsscheinwerfer des Thunderbird

Die ausschwenkbare Lenksäule des Thunderbird

Thunderbird, 2türiges Hardtop

Deutscher Werbeprospekt für den Ford T5, wie der Mustang hierzulande genannt wurde.
German advertising brochure for the Ford T5 as the Mustang was called in Germany.

Der Boss 429 war der größte Mustang-Motor. Hier wird er gerade in einen Wagen eingebaut.
The Boss 429 was the biggest Mustang engine. Here it is being placed in a car.

Der Ford Mustang begründete 1964 die neue Automobilklasse der sogenannten Pony Cars. Vergleichsweise viel Leistung für wenig Geld machte ihn vor allem für junge Käufer attraktiv.

In 1964, the Ford Mustang introduced the new "pony car" class of automobiles. With its relatively high power for little money, it was especially attractive for young customers.

Die Alpine A110 besaß einen Zentralrohrrahmen mit einer Karosserie aus Kunststoff.
The Alpine A110 had a central tube frame and an auto body made of plastic.

Alpine A110

Fahrzeugdaten

Hersteller:	Alpine
Land:	Frankreich
Modell:	A110
Bauzeit:	1962–1977
Länge:	3.850 mm
Breite:	1.460 mm
Höhe:	1.130 mm
Radstand:	2.100 mm
Leergewicht:	610 kg
Antriebsart:	Hinterrad
Höchstgeschwindigkeit:	175 km/h
Verbrauch:	ca. 9–12 Liter/100 km

*Werte für 1964 Alpine A110 1100

Der Renault-Händler und Rallye-Pilot Jean Rédélé gründete 1955 in seiner Heimatstadt Dieppe die Firma Alpine. Das Unternehmen baute Sportwagen auf Renault-Basis, die durch Erfolge im Motorsport schnell bekannt wurden.

1962 präsentierte Alpine das Modell A110: Das nur 1,13 Meter hohe Coupé mit Heckmotor wog lediglich 610 Kilogramm, verfügte über ein modifiziertes 1,0-Liter-Renault-Triebwerk mit 38 kW (52 PS) und erreichte eine Höchstgeschwindigkeit von 170 km/h. Das geringe Gewicht verdankte die Alpine ihrer leichten Karosserie aus Kunststoff, die 27 Kilogramm wog. Das Chassis bestand aus einem längs liegenden Zentralrohr mit Aufnahmen für Vorder- und Hinterachse plus Hilfsrahmen für Motor, Getriebe und Differenzial.

Die A110 wurde stetig weiterentwickelt und sorgte mit Erfolgen im Rallye-Sport für Furore. Ab 1973 besaß Renault die Aktienmehrheit von Alpine. Erst im 1977 endete die Produktion. Insgesamt entstanden 7.489 Exemplare des Modells A110.

Specifications

Manufacturer:	Alpine
Country:	France
Model:	A110
Produced:	1962–1977
Length:	3850 mm
Width:	1460 mm
Height:	1130 mm
Wheelbase:	2100 mm
Empty weight:	610 kg
Type of drive:	Rear-wheel drive
Max. speed:	175 km/h
Fuel consumption:	approx. 9–12 l/100 km

* data for the 1964 Alpine A110 1100

In 1955, Renault dealer and rally driver Jean Rédélé founded the Alpine company in his hometown, Dieppe. The company built sports cars based on Renault models. Due to their success in motorsports, these cars quickly became well known.

In 1962, Alpine introduced the A110 model: this rear-engine coupé with a height of no more than 1.13 meters weighed only 610 kg, used a modified 1.0-liter Renault engine with 38 kW (52 hp) und achieved a top speed of 170 km/h. The low weight of the Alpine was due to the light auto body, which was made of plastic and weighed only 27 kg. The chassis consisted of a longitudinal central tube with receptables for the front and rear axles plus an auxiliary frame for engine, transmission and differential.

The A110 was continually enhanced and caused sensations by its success in rally sports. From 1973 on, Renault owned the majority of Alpine shares. Production ended in 1977. In total, 7489 units of the model A110 were built.

Der Motor

Motordaten

Bauart:	Viertakt-Reihenmotor
Zylinderzahl:	4
Hubraum:	1.108 cm³
Bohrung:	70 mm
Hub:	72 mm
Leistung:	48 kW (65 PS)
	bei 6.500/min

In der ersten Version der Alpine A110 griff das Team von Firmenchef Jean Rédélé auf den Motor des Renault 8 zurück. Das 1,0-Liter-Triebwerk brachte es im R8 auf 32 kW (44 PS), doch für den Einsatz im neuen Sportwagen steigerten die Alpine-Techniker die Leistung des Vierzylinders auf 52 Pferdestärken (38 kW) bei 5.200 Umdrehungen pro Minute. Aber dies sollte erst der Anfang sein, denn im Heck der Alpine A110 saßen bald immer größere und stärkere Antriebsaggregate. Den Anfang machte 1964 eine 1,1-Liter-Version aus dem Renault 8 Major. Und schon 1965 brachte es der auf 1.296 Kubikzentimeter gewachsene Motor auf 88 kW (120 PS). Das reichte für eine Höchstgeschwindigkeit von unglaublichen 228 km/h. Damit war der Motor aus dem R8 aber an seine Grenze gekommen. Für mehr Hubraum sorgte ab 1966 der Vierzylinder aus dem R16. Für den Einsatz im Motorsport wurde der Motor auf 1,8 Liter Hubraum aufgebohrt. Diese Motorvariante leistete bis zu 132 kW (180 PS).

The engine

Engine specifications

Type:	Four-stroke in-line engine
Number of cylinders:	4
Displacement:	1108 cm³
Bore:	70 mm
Stroke:	72 mm
Power:	48 kW (65 hp) at 6500/min

For the first version of the Alpine A110, the team of company founder Jean Rédélé used the engine of the Renault 8. In the R8, this 1.0-liter engine provided 32 kW (44 hp), but for the deployment in the new sports car, the Alpine engineers increased the power of the 4-cylinder engine to 52 hp (38 kW) at 5200 rpm. But this was just the beginning! Soon, even bigger and more powerful engines were placed in the rear end of the Alpine A110. The first one was the 1.1-liter version from the Renault 8 Major in 1964. In 1965, the engine already had an increased displacement of 1296 cm³ and a power of 88 kW (120 hp), which allowed for a top speed of incredible 228 km/h. However, the R8 engine had thus reached its limits. From 1966 on, the new 4-cylinder engine of the R16 provided more displacement. For the use in motorsports, it was increased to a displacement of 1.8 liters. This version provided up to 132 kW (180 hp).

Für harte Rallye-Einsätze mussten Motor, Getriebe und Fahrwerk einer Alpine A110 stets penibel vorbereitet werden.

For tough rallies, engine, transmission and undercarriage of the Alpine A110 had to be prepared meticulously.

Der wassergekühlte Vierzylinder war in der Alpine A110 als Heckmotor installiert.
The water-cooled 4-cylinder engine was mounted in the rear of the Alpine A110.

Die Alpine A110 wurde durch Einsätze im Rallye-Sport legendär. So gewann der französische Sportwagen 1973 allein sechs WM-Läufe und bescherte Alpine-Renault den Marken-Titel der Rallye-Weltmeisterschaft.

Based on its success in rally sports, the Alpine A110 became a legend. The French sports car won six world championships in 1973 alone, which brought Alpine-Renault the brand title of the rally world championship.

Rund 2,8 Millionen Exemplare des Trabant 601 entstanden beim VEB Sachsenring in Zwickau.
Approx. 2.8 million Trabant 601 were built at VEB Sachsenring in Zwickau.

Trabant 601

Fahrzeugdaten

Hersteller:	VEB Sachsenring
Land:	DDR
Modell:	Trabant 601
Bauzeit:	1964–1990
Länge:	3.555 mm
Breite:	1.500 mm
Höhe:	1.440 mm
Radstand:	2.020 mm
Leergewicht:	615 kg
Antriebsart:	Vorderrad
Höchstgeschwindigkeit:	100 km/h
Verbrauch:	6,8 Liter/100 km

*Werte für 1964 Trabant 601

Der Trabant 601 war das meistgebaute Auto der DDR. Bei den VEB Sachsenring Automobil-werken entstanden zwischen 1964 und 1990 insgesamt über 2,8 Millionen Exemplare des Kleinwagens, die vor allem für den Binnenmarkt vorgesehen waren. In den früheren Werken der Auto Union im sächsischen Zwickau lief ab 1958 die erste Generation des Trabants unter der Bezeichnung „P50" vom Band. Er verfügte über einen luftgekühlten Zweizylinder-Zweitaktmotor, Frontantrieb und eine Karosserie aus Kunststoff. Diese Konstruktionsmerkmale fanden sich auch bei den Nachfolgemodellen. Der Trabant 601 leistete bei seiner Premiere 1964 17 kW (23 PS). Ab 1969 stieg die Leistung auf 19 kW (26 PS). Damit erreichte der 615 Kilogramm schwere Wagen eine Höchstgeschwindigkeit von 107 km/h. Neben der Limousine baute der VEB Sachsen-ring ab September 1965 auch eine Kombi-Version mit Namen „Universal". DDR-Bürger mussten bis zu 15 Jahre auf die Lieferung eines bestellten Trabants warten.

Specifications

Manufacturer:	VEB Sachsenring
Country:	GDR
Model:	Trabant 601
Produced:	1964–1990
Length:	3555 mm
Width:	1500 mm
Height:	1440 mm
Wheelbase:	2020 mm
Empty weight:	615 kg
Type of drive:	Front-wheel drive
Max. speed:	100 km/h
Fuel consumption:	6.8 l/100 km

* data for the 1964 Trabant 601

The Trabant 601 was the most produced car of the GDR. At VEB Sachsenring Automobilwerke, more than 2.8 million units of the subcompact car where built between 1964 and 1990, mostly for the home market. Beginning with 1958, the first generation of the Trabant with the desig-nation "P50" was manufactured in the former Auto Union factory in Zwickau, Saxony. It had an air-cooled 2-cylinder 2-stroke engine, front-wheel drive and a plastic auto body. These features were also adopted for the successor models. At its debut in 1964, the Trabant 601 provided 17 kW (23 hp). Beginning with 1969, the power was increased to 19 kW (26 hp), which gave the car of 615 kg a top speed of 107 km/h. In addition to the sedan, VEB Sachsenring also built a station wagon version called "Universal" from September 1965 on. GDR citizens had to wait up to 15 years for shipment of a Trabant that they had ordered.

Der Motor

Motordaten

Bauart:	Zweitakt-Reihenmotor
Zylinderzahl:	2
Hubraum:	594 cm³
Bohrung:	72 mm
Hub:	73 mm
Leistung:	17 kW (23 PS)
	bei 3.900/min

Der „Ost-Volkswagen" aus Zwickau besaß einen Zweizylinder-Zweitakter mit Einlass-Drehschieber. Das luftgekühlte Antriebsaggregat des Trabant 601 hatte einen Hubraum von 594 cm³. Bei Produktionsstart 1964 leistete dieser Motor 17 kW (23 PS) bei 3.900 Umdrehungen. Die Kurbelwelle war dreifach gelagert. Ein Horizontalvergaser war für die Gemischbildung zuständig.

Das Triebwerk war vorne quer im Trabant montiert. Über ein synchronisiertes Viergang-Getriebe wurden die Vorderräder angetrieben. Ab 1969 leistete der Zweizylinder 19 kW (26 PS) bei 4.200/min, doch eine grundlegende Weiterentwicklung fand nicht statt. In den 1980er-Jahren war das Motorkonzept schließlich hoffnungslos veraltet. Als Ersatz sollte ein in VW-Lizenz gebauter 1,1-Liter-Vierzylinder-Viertakter dienen. Ab Mai 1990 trat der Trabant 1.1 die Nachfolge des 601 an, doch in der DDR war der Wagen kaum noch verkäuflich. Insgesamt entstanden 39.474 Exemplare, bevor die Produktion im April 1991 endete.

The engine

Engine specifications

Type:	Two-stroke in-line engine
Number of cylinders:	2
Displacement:	594 cm³
Bore:	72 mm
Stroke:	73 mm
Power:	17 kW (23 hp)
	at 3900 rpm

The "people's car" from Zwickau had a 2-cylinder 2-stroke engine with a rotary intake valve. The air-cooled engine of the Trabant 601 had a displacement of 594 cm³. At production start in 1964, the engine provided 17 kW (23 hp) at 3900 rpm. The crankshaft was suspended at three points. A horizontal carburetor was used to provide the mixture.

The engine was mounted across in the front of the Trabant. The front wheels were driven via a synchronized four-gear transmission. From 1969 on, the 2-cylinder engine delivered 19 kW (26 hp) at 4200 rpm, but essential enhancements never took place. In the 1980s, the engine concept was outdated beyond hope. As a replacement, a 1.1-liter 4-cylinder 4-stroke engine was built under license by VW. The Trabant 1.1 succeeded the 601 beginning in May 1990; however, the car was no longer marketable in the GDR. In total, only 39,474 units were built until production ceased in April 1991.

DDR-Nostalgie: Ein Trabant 601 vor dem Karl-Marx-Denkmal in Chemnitz.

GDR nostalgia: a Trabant 601 in front of the Karl Marx monument in Chemnitz.

19 kW (26 PS) leistete der Zweizylinder-Zweitaktmotor des Trabant 601 ab 1969.
From 1969 on, the 2-cylinder 2-stroke engine of the Trabant 601 delivered 19 kW (26 hp).

Der VEB Sachsenring baute ab 1964 den Trabant 601 in den ehemaligen Werken der Auto Union im sächsischen Zwickau. Der DDR-Kleinwagen war mit einem Zweitaktmotor ausgestattet und lief bis 1990 vom Band.

Beginning in 1964, the VEB Sachsenring built the Trabant 601 in the former Auto Union factories in Zwickau, Saxony. The GDR subcompact car was equipped with a 2-stroke engine and produced until 1990.

Rüsselsheimer Top-Modell: Das Opel Diplomat A Coupé wurde von 1965 bis 1967 gebaut.

Top model from Rüsselsheim: the Opel Diplomat A Coupé was built from 1965 to 1967.

Opel Diplomat A

Fahrzeugdaten

Hersteller:	Opel
Land:	Deutschland
Modell:	Diplomat A
Bauzeit:	1964–1968
Länge:	4.948 mm
Breite:	1.902 mm
Höhe:	1.454 mm
Radstand:	2.845 mm
Leergewicht:	1.630 kg
Antriebsart:	Hinterrad
Höchstgeschwindigkeit:	200 km/h
Verbrauch:	15,3 Liter/100 km

*Werte für 1966 Opel Diplomat 5,4

1964 schlug bei Opel in Rüsselsheim die Geburtsstunde für drei neue Modelle: Kapitän, Admiral und Diplomat – gemeinsam kurz „KAD" genannt. Die Fahrzeuge der KAD-Baureihe verfügten über die gleiche Karosserie, unterschieden sich aber in Ausstattung und Motorisierung.

Der Diplomat A war das damalige Top-Modell der deutschen Tochter von General Motors. Serienmäßig besaß er einen 4,6-Liter-V8-Motor und ein Automatikgetriebe. Beides bezogen die Hessen von der Konzernmutter aus den USA. Der Opel Diplomat konnte bei seinem Erscheinen mit Komfort punkten. So umfasste bereits die Serienausstattung eine Servolenkung und elektrische Fensterheber – in den 1960er-Jahren noch eine absolute Ausnahme auf dem deutschen Markt.

Von 1965 an baute Karmann in Osnabrück für Opel das Diplomat V8 Coupé. Dieses Modell war mit einem 5,4-Liter-V8-Triebwerk ausgestattet, das 169 kW (230 PS) leistete. Ab 1966 war dieser Motor auf Wunsch auch für die Limousine lieferbar.

Specifications

Manufacturer:	Opel
Country:	Germany
Model:	Diplomat A
Produced:	1964–1968
Length:	4948 mm
Width:	1902 mm
Height:	1454 mm
Wheelbase:	2845 mm
Empty weight:	1630 kg
Type of drive:	Rear-wheel drive
Max. speed:	200 km/h
Fuel consumption:	15.3 l/100 km

* data for the 1966 Opel Diplomat 5.4

In 1964, three new models were born at the Opel factories in Rüsselsheim, namely the Kapitän, the Admiral and the Diplomat, collectively called "KAD". The vehicles of the KAD series featured the same auto body but had different engines and interiors.

The Diplomat A was the then top model of the German General Motors subsidiary. It had a 4.6-liter V8 engine and an automatic transmission in base, both sourced from the US parent company. The Opel Diplomat heavily scored with comfort. The standard equipment contained power steering and electrical window lifters. In the 1960s, this was an absolute rarity on the German market.

Beginning in 1965, Karmann in Osnabrück built the Diplomat V8 Coupé for Opel. This model had a 5.4-liter V8 engine with 169 kW (230 hp). From 1966 on, this engine was optionally available for the sedan.

Der Motor

Motordaten

Bauart:	Viertakt-90-Grad-V-Motor
Zylinderzahl:	8
Hubraum:	5.354 cm³
Bohrung:	101,6 mm
Hub:	82,6 mm
Leistung:	169 kW (230 PS)
	bei 4.700/min

Der V8-Motor im Opel Diplomat A war ein alter Bekannter aus den USA. Dort verbaute Chevrolet das Triebwerk millionenfach. Dieser sogenannte Small Block mit unten liegender Nockenwelle verfügte über einen Hubraum von 4.638 cm³ und leistete 1964 im Opel Diplomat 139 kW (190 PS) bei 4.600 Umdrehungen. Motorblock und Zylinderkopf bestanden aus Grauguss. Noch mehr Leistung bot ab 1965 eine 5,4-Liter-Version des V8-Triebwerks. Zunächst war dieser Motor nur im neuen Diplomat V8 Coupé serienmäßig verbaut, bevor das Aggregat ab 1966 gegen Aufpreis auch für die Limousine angeboten wurde. Die Top-Motorisierung brachte es auf 169 kW (230 PS) bei 4700/min. Das Drehmoment war bemerkenswert: Der Achtzylinder im Opel produzierte 435 Nm bei 3.000 Touren.

Der Diplomat war nur mit einer Powerglide-Zweigangautomatik lieferbar. Trotzdem waren die Fahrleistungen durchaus sportlich: Für den Sprint von 0 auf 100 km/h benötigte der Diplomat rund zehn Sekunden.

The engine

Engine specifications

Type:	Four-stroke 90° V engine
Number of cylinders:	8
Displacement:	5354 cm³
Bore:	101.6 mm
Stroke:	82.6 mm
Power:	169 kW (230 hp)
	at 4700 rpm

The V8 engine of the Opel Diplomat A was an old acquaintance from the USA where Chevrolet used it in millions. The so-called Small Block with a camshaft in bottom position had a displacement of 4638 cm³. In 1964, it provided 139 kW (190 hp) at 4600 rpm in the Opel Diplomat. Engine block and cylinder heads were made of grey cast iron. From 1965 on, a 5.4-liter version of the V8 engine provided even more power. At first, the new engine was only mounted in the new Diplomat V8 Coupé; however, beginning in 1966, it was also available for the sedan at an additional price. This top engine provided 169 kW (230 hp) at 4700 rpm. The torque of 435 Nm at 3000 rpm was also impressive. The Diplomat was only available with a Powerglide two-gear automatic transmission. Nonetheless the driving characteristics were quite sporty: the Diplomat needed approx. 10 seconds to accelerate from 0 to 100 km/h.

Der Opel Diplomat besaß einen V8-Motor aus dem Baukasten der Konzernmutter General Motors.
The Opel Diplomat had a V8 engine from the construction kit of the parent company General Motors.

Opel trat in den 1960er-Jahren mit V8-Technik aus den USA in der Oberklasse gegen die Konkurrenz aus Stuttgart an. Chevrolet lieferte den Motor für den Opel Diplomat A, der 1964 seine Premiere feierte.

In the 1960s, Opel used V8 technology from the USA to compete in the upper class segment against the competition from Stuttgart. The engine for the Open Diplomat A, which debuted in 1964, was supplied by Chevrolet.

Der offene Fiat 124 Spider kam im Jahr 1966 auf den Markt und wurde bis 1985 produziert.
The open Fiat 124 Spider was launched in 1966 and produced until 1985.

Fiat 124 Spider

1966

Fahrzeugdaten

Hersteller:	Fiat
Land:	Italien
Modell:	124 Spider
Bauzeit:	1966–1985
Länge:	3.970 mm
Breite:	1.610 mm
Höhe:	1.250 mm
Radstand:	2.280 mm
Leergewicht:	945 kg
Antriebsart:	Hinterrad
Höchstgeschwindigkeit:	170 km/h
Verbrauch:	10,5 Liter/100 km

*Werte für 1966 Fiat 124 Spider

Der 1966 vorgestellte Fiat 124 Spider war erkennbar als direkter Konkurrent des fast zeitgleich präsentierten Alfa Romeo Spider konzipiert. Beide Entwürfe stammten aus der Feder von Designer Sergio Pininfarina – und auch im Grundkonzept ähnelten sich die italienischen Rivalen.

Ab 1966 war der Fiat zunächst mit einem 1,4-Liter-Vierzylinder-Motor ausgestattet, der 67 kW (90 PS) bei sportlichen 6.500/min leistete. Das reichte für eine Höchstgeschwindigkeit von 170 km/h.

Hubraum und Leistung wuchsen in den folgenden Jahren. Gebaut wurde der 124 Spider bei Pininfarina in Turin. Als Basismodell für den Einsatz im Motorsport entstand 1972 eine bei Abarth entwickelte Version mit 94 kW (128 PS). In der Rallye-WM feierte der 124 große Erfolge.

Als Fiat den Spider 1982 offiziell aus dem Programm nahm, ging die Produktion weiter – und der Wagen wurde als Pininfarina Spidereuropa angeboten. Zwischen 1966 und 1985 liefen insgesamt rund 198.000 Exemplare vom Band.

Specifications

Manufacturer:	Fiat
Country:	Italy
Model:	124 Spider
Produced:	1966–1985
Length:	3970 mm
Width:	1610 mm
Height:	1250 mm
Wheelbase:	2280 mm
Empty weight:	945 kg
Type of drive:	Rear-wheel drive
Max. speed:	170 km/h
Fuel consumption:	10.5 l/100 km

* data for the 1966 Fiat 124 Spider

The Fiat 124 Spider introduced in 1966 was evidently designed as a direct competitor to the Alfa Romeo Spider, presented nearly at the same time. Both constructions were the brainchild of designer Sergio Pininfarina. Even the basic concepts of the Italian rivals were similar.

At first, the Fiat was equipped with a 1.4-liter 4-cylinder engine, which provided 67 kW (90 hp) at a sporty 6500 rpm. This was sufficient to reach a top speed of 170 km/h. In the following years, displacement and engine power increased. The 124 Spider was built at Pininfarina in Turin. In 1972, a base model for motorsports with 94 kW (128 hp) was developed by Abarth. The 124 was very successful in the rally world championship

Fiat phased out the Spider in 1982, but the production continued nonetheless. The car was now sold as Pininfarina Spidereuropa. Between 1966 and 1985, approx. 198,000 units in total left the assembly lines.

Der Motor

Motordaten

Bauart:	Viertakt-Reihenmotor
Zylinderzahl:	4
Hubraum:	1.438 cm³
Bohrung:	80 mm
Hub:	71,5 mm
Leistung:	67 kW (90 PS)
	bei 6.500/min

Das Cabrio aus dem Hause Fiat kam 1966 mit einem sehr drehfreudigen Motor auf den Markt: Der 1,4-Liter große Reihen-Vierzylinder verfügte über zwei oben liegende Nockenwellen, zwei Ventile pro Zylinder und leistete 67 kW (90 PS) bei 6.500/min. Das Triebwerk besaß einen Graugussblock, während der Zylinderkopf aus Aluminium gefertigt war. Das maximale Drehmoment des wassergekühlten Motors lag bei 108 Nm, das er bei 3.600 Umdrehungen pro Minute produzierte. Ein Vergaser aus dem Hause Weber sorgte für das passende Kraftstoff-Luft-Gemisch.

Die Fiat-Motorenentwickler vergrößerten den Hubraum des Cabrios bis 1978 in mehreren Schritten schließlich auf 2,0 Liter. Zwischen 1983 und 1985 entstand mit dem Pininfarina Spidereuropa Volumex die leistungsstärkste Variante des Fiat 124 Spider: Ausgestattet mit einem Roots-Kompressor leistete der Vierzylinder-Motor 99 kW (135 PS) bei 5600/min. Von diesem Modell produzierte Pininfarina insgesamt nur 500 Exemplare.

The engine

Engine specifications

Type:	Four-stroke in-line engine
Number of cylinders:	4
Displacement:	1438 cm³
Bore:	80 mm
Stroke:	71,5 mm
Power:	67 kW (90 hp)
	at 6500 rpm

The Fiat cabriolet launched in 1966 had a very speedy engine. The 1.4-liter 4-cylinder in-line engine had two overhead camshafts and two valves per cylinder and provided 67 kW (90 hp) at 6500 rpm. The block was made of grey cast iron, while the cylinder head was made of aluminum. The maximum torque of the water-cooled engine amounted to 108 Nm and was reached at 3600 rpm. A Weber carburetor provided the suitable fuel-air mixture.

Until 1978, the Fiat engine developers increased the displacement of the cabriolet step by step up to 2.0 liters. Between 1983 and 1985, the most powerful version of the Fiat 124 Spider was created, namely the Pininfarina Spidereuropa Volumex. Thanks to a Roots compressor, the 4-cylinder engine provided 99 kW (135 hp) at 5600 rpm. Pininfarina only built 500 units of this model.

Einheit: Der 1,4-Liter-Vierzylinder war im 124 Spider mit einem Fünfgang-Getriebe verbunden.
The 1.4-liter 4-cylinder engine of the 124 Spider was connected to a five-gear transmission.

Mit dem Fiat 124 Spider gelang Designer Sergio Pininfarina 1966 ein großer Wurf. Eine Abarth-Version machte zudem in den 1970er-Jahren Schlagzeilen. Erst 1985 endete die Produktion des offenen Zweisitzers.

With the design of the Fiat 124 Spider in 1966, Sergio Pininfarina hit the jackpot. An Abarth version made a splash in the 1970s. Production of the open two-seater ended not before 1985.

Der Subaru 1000 war der erste mit einem Boxermotor ausgestattete Wagen des japanischen Herstellers.
The Subaru 1000 was the first car of the Japanese manufacturer that sported a flat engine.

Subaru 1000

Fahrzeugdaten

Hersteller:	Subaru
Land:	Japan
Modell:	1000
Bauzeit:	1966–1969
Länge:	3.930 mm
Breite:	1.480 mm
Höhe:	1.390 mm
Radstand:	2.420 mm
Leergewicht:	670 kg
Antriebsart:	Vorderrad
Höchstgeschwindigkeit:	ca. 135 km/h
Verbrauch:	5 Liter/100 km

Der 1966 vorgestellte Subaru 1000 war in der unteren Mittelklasse angesiedelt und ausschließlich für den japanischen Markt bestimmt. Er war der erste Kleinwagen aus dem Hause Subaru, das zuvor nur Kleinstwagen produziert hatte. Das Modell 1000 wurde von 1966 bis 1969 gebaut. Es handelte sich um das erste Personenauto mit Frontantrieb und Boxermotor des japanischen Herstellers. Anfangs wurde der Subaru 1000 nur als Limousine angeboten. Das zweitürige Coupe und ein viertüriger Kombi folgten jedoch im Jahr 1967. In Japan wurden über 4.000 Einheiten des Subaru 1000 verkauft. Damit war der Wagen für sich betrachtet nicht gerade ein wirtschaftlicher Erfolg. Bei Subaru läutete er allerdings die lange Ära des Boxermotorenbaus ein. Bis 2016 stellte Subaru über 16 Millionen dieser flachen, sehr laufruhigen und überaus kompakt gebauten Aggregate her und baute diese in seine Fahrzeuge ein. Heute stattet der Hersteller seine Pkw ausschließlich mit Boxermotoren aus.

Specifications

Manufacturer:	Subaru
Country:	Japan
Model:	1000
Produced:	1966–1969
Length:	3930 mm
Width:	1480 mm
Height:	1390 mm
Wheelbase:	2420 mm
Empty weight:	670 kg
Type of drive:	Front-wheel drive
Max. speed:	approx. 135 km/h
Fuel consumption:	5 l/100 km

The Subaru 1000 introduced in 1966 was a car of the lower middle class and exclusively intended for the Japanese market. It was the first subcompact car by Subaru, which had previously only produced micro cars. The model 1000 was built from 1966 to 1969. It was the first passenger car of the Japanese manufacturer with front-wheel drive and flat engine. First, the Subaru 1000 was only available as sedan, but a two-door coupé and a four-door station wagon already followed in 1967. More than 4000 Subaru 1000 were sold in Japan. Taken by itself, the car was thus not a huge economic success. However, it heralded the long era of flat-engine construction at Subaru. Until 2016, Subaru had built more than 16 million of these compact and smooth-running engines and mounted them in its cars. Today, this manufacturer equips its own passenger cars exclusively with flat engines.

Der Motor

Motordaten

Bauart:	Viertakt-Boxermotor
Zylinderzahl:	4
Hubraum:	977,2 cm³
Bohrung:	72 mm
Hub:	60 mm
Leistung:	40 kW (55 PS)
	bei 6.000/min

In den Subaru 1000 wurde ein für damalige Verhältnisse einzigartiger Boxermotor mit Wasserkühlung eingebaut. Er trug die Typenbezeichnung „EA-52". Durch die Wasserkühlung konnte die Geräuschentwicklung des Aggregats vermindert werden, da der bei luftgekühlten Motoren übliche Lüfter entfiel. Der Motor war mit je einer oben liegenden Nockenwelle pro Zylinderbank ausgestattet. Das Aggregat gab bei 6.000 Touren 40 kW (55 PS) ab. Das Drehmoment lag bei 76 Nm, die bei 3.200 Umdrehungen pro Minute zur Verfügung standen. Das Kurbelgehäuse und die Zylinderköpfe des Vierzylinder-Boxermotors waren aus Aluminium gefertigt. Dieses kostete in den 1960er-Jahren zwar das 14-Fache billigen Graugusses, half aber, das Gewicht der Maschine um 15 Prozent zu verringern. So leistete der Motor seinen Beitrag dazu, dass der Subaru 1000 nicht einmal 700 Kilogramm auf die Waage brachte und mit einem Verbrauch von nur 5 Litern auf 100 Kilometer zu den sparsamsten Fahrzeugen seiner Zeit zählte.

The engine

Engine specifications

Type:	Four-stroke flat engine
Number of cylinders:	4
Displacement:	977.2 cm³
Bore:	72 mm
Stroke:	60 mm
Power:	40 kW (55 hp)
	at 6000 rpm

The Subaru 1000 had a water-cooled flat engine, which was unique for its time. The water-cooling helped to reduce the engine noise because the fan required for an air-cooled engine was no longer needed. The engine with the type designation "EA-52" had one overhead camshaft per cylinder bank and produced 40 kW (55 hp) at 6000 rpm. The maximum torque of 76 Nm was available at 3200 rpm. Crankcase and cylinder heads of the 4-cylinder flat engine were made of aluminum. In the 1960s, this material was 14 times more expensive than grey cast iron, but it helped to reduce the weight of the engine by 15 %. In this way, the engine contributed to the low weight of less than 700 kg and thus to a fuel consumption of only 5 l/100 km, which made the Subaru 1000 to one of the most fuel-saving vehicles of its time.

Der Subaru-Boxermotor EA-52 läutete 1966 die Tradition der Boxermotoren bei Subaru ein.
In 1966, the Subaru flat engine EA-52 heralded the flat engine tradition of Subaru.

Mit wassergekühltem Boxermotor und Vorderradantrieb war der lediglich in Japan verkaufte Subaru 1000 mit einer für das Unternehmen zukunftsweisenden Technik ausgestattet.

With a water-cooled flat engine and front-wheel drive, the Subaru 1000, which was only sold in Japan, was equipped with advanced technology and should determine the future course of the company.

Der mit einem 4,9-Liter-V8 ausgestattete Chevrolet Camaro Z-28 wurde 1967 nur 609 Mal gebaut.
The Chevrolet Camaro Z-28 had a 4.9-liter V8 engine. Only 609 units were built in 1967.

Chevrolet Camaro Z-28

Fahrzeugdaten

Hersteller:	Chevrolet
Land:	USA
Modell:	Camaro Z-28
Bauzeit:	1967–1969
Länge:	4.691 mm
Breite:	1.842 mm
Höhe:	1.306 mm
Radstand:	2.743 mm
Antriebsart:	Hinterrad
Höchstgeschwindigkeit:	210 km/h
Verbrauch:	ca. 20 Liter/100 km

Nachdem Ford von 1964 an mit dem Mustang äußerst erfolgreich war, wollte Chevrolet das neue Segment der Pony Cars nicht kampflos der Konkurrenz überlassen und schuf 1966 den Camaro als Coupé sowie als Cabriolet. Bereits im ersten Jahr wurden davon mehr als 220.000 Einheiten an den Mann gebracht. Ein Mythos war geboren. Die erste bis 1970 gebaute Generation des Camaro nutzte die seinerzeit neu entwickelte Chevrolet-Nova-Plattform von 1966, welche die Bodengruppe von der A-Säule bis zum Heck umfasste. Neben Sechszylinder-Reihenmotoren sorgten vor allem V8-Triebwerke für Vortrieb. Im Dezember 1966 debütierte der Camaro in der Version Z-28, die zunächst unbeworben blieb. Mit dem 1967 nur 609 Mal gebauten Z-28 wollte man eine erfolgreiche Teilnahme an der „Sports Car Club of America Trans Am"-Rennserie für straßenzugelassene Fahrzeuge erreichen, was auch gelang. Bis 2002 entstanden vier Camaro-Modellgenerationen, und 2009 ließ General Motors ihn mit einer fünften Generation wieder auferstehen.

Specifications

Manufacturer:	Chevrolet
Country:	USA
Model:	Camaro Z-28
Produced:	1967–1969
Length:	4691 mm
Width:	1842 mm
Height:	1306 mm
Wheelbase:	2743 mm
Type of drive:	Rear-wheel drive
Max. speed:	210 km/h
Fuel consumption:	approx. 20 l/100 km

When Ford successfully launched Mustang in 1964, Chevrolet did not want to leave the new pony car segment to the competitors and thus came up with the Camaro in 1966, which was built as a coupé and a convertible. In the first year alone, more than 220,000 units were sold. A new legend was born. The first generation of the Camaro – built until 1970 – used the newly developed Chevrolet Nova platform of 1966, which contained the floor assembly from the A-pillar to the tail. Apart from the 6-cylinder in-line engine, propulsion was provided in particular by a V8 engine. In December 1966, the Camaro Z-28 debuted. At first, it was not advertised. Only 609 units of the Z-28 were built in 1967. Chevrolet wanted to use it for successfully participating in the Sports Car Club of America Trans-Am races for homologated vehicles. Up to 2002, there were four Camaro generations. In 2009, General Motors revived it with a fifth generation.

Der Motor

Motordaten

Bauart:	Viertakt-90-Grad-V-Motor
Zylinderzahl:	8
Hubraum:	4.942 cm³
Bohrung:	101,6 mm
Hub:	76,2 mm
Leistung:	216 kW (290 PS)
	bei 5.800/min

Der Chevrolet Camaro bot eine Reihe von Sechs- und Achtzylinder-Motoren mit Leistungen von 103 kW (140 PS) bis 276 kW (375 PS). Der 90-Grad-V8-Motor des Camaro Z-28 von 1967 ist auch als 302-Motor bekannt geworden, was sich auf seinen Hubraum von 302 Cubic Inch bezog. Er wurde für die Trans-Am-Rennserie des Sports Car Club of America (SCCA) und das dort geltende Hubraumlimit von 305 Cubic Inch entwickelt. Dafür kombinierte man den Block eines 327-Motors mit der Kurbelwelle eines 287-Triebwerks, was zu exakt 302,3 Cubic Inch Hubvolumen führte. Der 4,9-Liter-V8-Motor reichte sein Drehmoment an ein 4-Gang-Schaltgetriebe der Firma Muncie weiter und leistete 216 kW (290 PS) bei 5.800 Umdrehungen pro Minute. Der Motor galt aufgrund seiner Charakteristik als durchaus anspruchsvoll zu fahren, da er vergleichsweise hohe Drehzahlen bis zu 7.500 Touren ermöglichte. Von kundiger Hand bewegt, sorgte das Triebwerk jedoch für zahlreiche Rennsiege des Camaro Z-28.

The engine

Engine specifications

Type:	4-stroke 90° V-engine
Number of cylinders:	8
Displacement:	4942 cm³
Bore:	101.6 mm
Stroke:	76.2 mm
Engine power:	216 kW (290 hp)
	at 5800 rpm

The Chevrolet Camaro was built with various 6- and 8-cylinder engines in the range of 103 kW (140 hp) to 276 kW (375 hp). The 90° V8 engine of the Camaro Z-28 of 1967 was also called the 302 engine because of its displacement of 302 cubic inches. It had been developed to comply to the 305 cubic inch displacement limit of the Trans-Am races of the Sports Car Club of America (SCCA). To this end, the block of a 327 engine was combined with the crankshaft of a 287 engine, which resulted in a displacement of exactly 302.3 cubic inches. The 4.9-liter V8 engine transferred its torque to a four-gear manual transmission by Muncie and provided 216 kW (290 hp) at 5800 rpm. Because of its relatively high speed of up to 7500 rpm, the engine was said to be rather challenging to drive. However, controlled with a knowing hand, the engine facilitated numerous racing wins for the Camaro Z-28.

Der 4,9-Liter-V8 des Camaro Z-28 von 1967 leistete 216 kW (290 PS) bei 5.800 Touren.
The 4.9-liter V8 engine of the Camaro Z-28 of 1967 provided 216 kW (290 hp) at 5800 rpm.

Mit dem Camaro Z-28 und seinem 216 kW (290 PS) starken 302-Cubic-Inch-Motor schuf Chevrolet ein potentes Auto für die Trans-Am-Rennserie des Sports Car Club of America (SCCA) mit 305-Cubic-Inch-Hubraumlimit.

With the Camaro Z-28 and its 302 cubic inch engine with 216 kW (290 hp), Chevrolet created a potent car for the Trans-Am races of the Sports Car Club of America (SCCA) with a 305 cubic inch displacement limit.

Der NSU Ro 80 nahm formensprachlich ganze Generationen von Automobilen vorweg.
The design of the NSU Ro 80 anticipated many future automobile generations.

NSU Ro 80

Fahrzeugdaten

Hersteller:	NSU
Land:	Deutschland
Modell:	Ro 80
Bauzeit:	1967–1977
Länge:	4.780 mm
Breite:	1.760 mm
Höhe:	1.410 mm
Radstand:	2.860 mm
Leergewicht:	1.280 kg
Antriebsart:	Vorderrad
Höchstgeschwindigkeit:	ca. 180 km/h
Verbrauch:	11,2 Liter/100 km

Als NSU 1967 den Ro 80 vorstellte, kam er einer formensprachlichen Revolution gleich – denn er nahm viele Designmerkmale von Autos der 1970er- bis weit in die 1980er-Jahre vorweg. Eine derart schlanke, elegante, vor allem aber aerodynamische Karosserie war bis dahin ohne Beispiel. Getrost kann man sie sogar als Urform ganzer Audi-Modellgenerationen betrachten. Zu den positiven Fahreigenschaften des Fahrzeugs trug auch die umfassende Serienausstattung bei, die unter anderem vier Scheibenbremsen, ein Zweikreis-Bremssystem sowie eine Dreigang-Kupplungsautomatik beinhaltete. Zudem beeindruckte die geräumige Fahrgastzelle, die vorne wie hinten reichlich Beinfreiheit bot. Weiterhin ließ sich der Kofferraum durch getrennt herausnehmbare Rückenlehnen vergrößern. Der Ro 80 wurde von 1967 bis 1977 gebaut. Wegen technischer Probleme mit dem Wankelmotor haftete ihm bald ein negativer Ruf an, der viele potenzielle Käufer abschreckte. So verließen nur 37.398 Einheiten die Werkshallen.

Specifications

Manufacturer:	NSU
Country:	Germany
Model:	Ro 80
Produced:	1967–1977
Length:	4780 mm
Width:	1760 mm
Height:	1410 mm
Wheelbase:	2860 mm
Empty weight:	1280 kg
Type of drive:	Front-wheel drive
Max. speed:	approx. 180 km/h
Fuel consumption:	11.2 l/100 km

The introduction of the NSU Ro 80 in 1967 was tantamount to a design revolution, as this car already anticipated many design elements of cars that would be built in the 1970s and 1980s. Its slim, elegant and above all aerodynamic auto body was unprecedented. It could arguably be called the original mould of many Audi model generations. The comprehensive base equipment contributed to the positive driving characteristics. Among others it encompassed four disk brakes, a dual circuit braking system and a three-gear automatic clutch. It also sported an impressive and spacious passenger compartment with much legroom at the front and back seats. Furthermore, the trunk could be extended by taking out the separable seat backs. The Ro 80 was built from 1967 until 1977. Due to technical problems of the Wankel engine, it soon obtained a bad reputation, which scared off many potential customers. For this reason, only 37,398 units ever left the factories.

Der Motor

Motordaten

Bauart:	Zweikammer-Kreiskolben-Motor (Wankel)
Anz. Kammern:	2
Hubraum:	2 mal 497,5 cm³
Leistung:	84,6 kW (115 PS) 5.500/min

Anstatt eines üblichen Hubkolbenmotors baute NSU in den gehobenen Mittelklassewagen einen sogenannten Dreh- oder auch Kreiskolbenmotor ein – besser als „Wankelmotor" bekannt. Die Entwicklungsarbeit des neuen Motors verschlang Unsummen und setzte sich noch fort, nachdem der Ro 80 längst auf dem Markt war. Aufgrund eines Konstruktionsfehlers kam es vermehrt zu Motorschäden. Ihre Ursachen lagen in falschen Materialkombinationen sowie Problemen mit den Dichtleisten. Für Probleme sorgte ferner der Einsatz von zwei Zündkerzen pro Verbrennungskammer. Hier hatte man mit der Einstellung der Zündzeitpunkte sowie mit dem hohen Abbrand der Unterbrecherkontakte zu kämpfen. Es dauerte bis 1971, bis der Ro 80 störungsfrei lief. Der Ruf des Wagens war zu diesem Zeitpunkt jedoch schon zu sehr angeschlagen. Der Motor gab seine maximale Leistung von 84,6 kW (115 PS) bei 5.500 Touren ab. Es kamen zwei Flachstrom-Registervergaser mit Beschleunigerpumpen zum Einsatz.

The engine

Engine specifications

Type:	Twin-chamber rotary piston engine (Wankel engine)
Number of chambers:	2
Displacement:	2 x 497.5 cm³
Power:	84.6 kW (115 hp) at 5500 rpm

Instead of a usual piston engine, NSU equipped its upper middle class car with a so-called rotary piston engine or Wankel engine. The development of the new engine cost an enormous amount of money and even continued after the Ro 80 was already on the market. Because of a design flaw, engine faults occurred quite often. This was caused by an improper combination of materials and by problems with the sealing strips. Further problems arose out of the use of two spark plugs per combustion chamber, which made the engineers struggle with setting the ignition timing and with the high burn-off loss of the breaker contacts. It was no sooner than 1971, when the Ro 80 finally ran smoothly. At this time, however, the reputation of the car was already tarnished. The engine provided its maximum power of 84.6 kW (115 hp) at 5500 rpm. Two flat-stream staged carburetors with accelerator pumps were used.

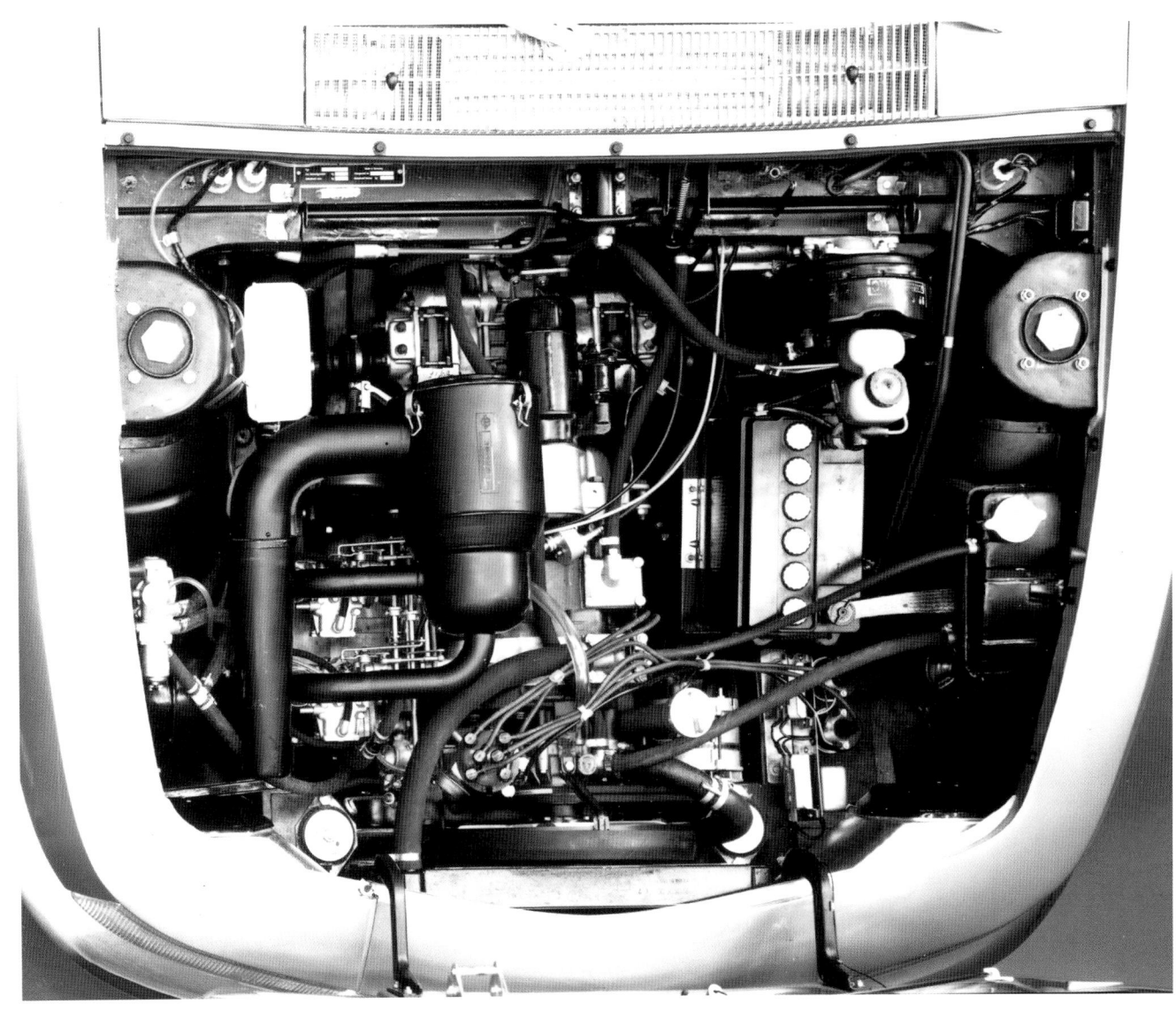

Der laufruhige und leistungsstarke Wankelmotor des NSU Ro 80 machte anfangs Probleme.
The smooth-running and powerful Wankel engine of the NSU Ro 80 caused severe problems in the beginning.

Der NSU Ro 80 nahm im Hinblick auf die Formensprache vieles im Automobilbau vorweg. Aufgrund von Motorproblemen war dem zukunftsweisenden
Konzept jedoch nur geringer Erfolg beschieden.

The NSU Ro 80 anticipated many future design elements. Due to engine
problems, however, the pioneering concept was not successful.

1968 erhielt der 110 S Cosmo Sport einen längeren Radstand und einen stärkeren Wankelmotor.
In 1968, the 110 S Cosmo Sport got a longer wheelbase and a more powerful Wankel engine.

Der Mazda 110 S Cosmo Sport war der erste japanische Wagen mit Wankelmotor.
The Mazda 110 S Cosmo Sport 110 was the first Japanese car with a Wankel engine.

Mazda 110 S Cosmo Sport

1967

Fahrzeugdaten

Hersteller:	Mazda
Land:	Japan
Modell:	110 S Cosmo Sport
Bauzeit:	1967–1972
Länge:	4.140 mm
Breite:	1.594 mm
Höhe:	1.156 mm
Radstand:	2.200 mm
Leergewicht:	930 kg
Antriebsart:	Hinterrad
Höchstgeschwindigkeit:	193 km/h
Verbrauch:	k.A.

Nachdem sich Mazda 1961 die Rechte am Wankelmotor von NSU gesichert hatte, begann man unverzüglich mit der Entwicklung eines eigenen Kreiskolben-Triebwerks. Fast gleichzeitig wurde das Cosmo-Projekt ins Leben gerufen. 1963 wurde der erste Prototyp des Cosmo Sport fertigge-stellt, und bereits 1966 konnte eine Vorserie von 80 Stück gefertigt werden. Die Wagen gingen an Fachhändler, welche die Akzeptanz der Kunden für das neue Sport-Coupé testen sollten. Der von 1967 bis 1972 gebaute Mazda 110 S Cosmo Sport bestach nicht nur durch sein futuristisches Aussehen. Er war auch weltweit der erste Serienwagen mit Zweischeiben-Wankel-motor – noch vor dem deutschen NSU Ro 80. Die erste Version von 1967 wurde nur 343 Mal gebaut. Sie ist auch als „L10A" bekannt. Vom L10B, der mit stärkerem Motor und verlängertem Radstand aufwartete, wurden von 1968 an 1.176 Einheiten hergestellt. Alle 110 S Cosmo Sport wurden ausschließlich in Japan verkauft. Es gab sie daher nur als Rechtslenker.

Specifications

Manufacturer:	Mazda
Country:	Japan
Model:	110 S Cosmo Sport
Produced:	1967–1972
Length:	4140 mm
Width:	1594 mm
Height:	1156 mm
Wheelbase:	2200 mm
Empty weight:	930 kg
Type of drive:	Rear-wheel drive
Max. speed:	193 km/h
Fuel consumption:	n.a.

After Mazda secured the rights to the NSU Wankel engine in 1961, the company immediately started to develop its own rotary piston engine. Nearly at the same time, the Cosmo project was launched. In 1963, the first prototype of the Cosmo Sport was completed, and in 1966, a pilot series of 80 units was produced. These cars were sent to retailers who should test the consumer acceptance of the new sports coupé. The Mazda 110 S Cosmo Sport was built from 1967 to 1972. It did not only impress by its futuristic appearance, but it was also the first production car in the world with a twin-rotor Wankel engine, even before the German NSU Ro 80. Of the first version of 1967, only 343 units were built. It is also known as "L10A." Since 1968, 1176 units of the L10B with a more powerful engine and longer wheelbase were manufactured. The 110 S Cosmo Sport was sold exclusively in Japan. For this reason, the model was only built as right-hand drive vehicles.

Der Motor

Motordaten

Bauart:	Kreiskolben-Motor (Wankel)
Kammerzahl:	2
Kammervolumen:	2 x 491 cm³
Leistung:	81 kW (110 PS)
	bei 7.000/min

Für die Entwicklung ihres Wankelmotors ließen sich die Mazda-Ingenieure Zeit. Das Resultat war ein Aggregat mit ausgezeichneten Laufeigenschaften und hoher Zuverlässigkeit. Gerade daran mangelte es vielen Wankelmotoren bisher. Häufig zu Ausfällen führten vor allem Dichtungsprobleme. Die aber hatten die Japaner von Beginn an im Griff. Nachdem der erste „L8A" genannte Prototyp ein Kammervolumen von zweimal 398 cm³ besaß, wurde es für den Serienmotor (L10A) auf zweimal 491 cm³ vergrößert. Weiter setzte man anstatt auf die Seiten- und Umfangseinlässe des L8A beim L10A nur noch auf Seiteneinlässe. Dies sorgte für ein verbessertes Drehmoment und bessere Laufeigenschaften im unteren Drehzahlbereich. Der Zweischeiben-Kreiskolbenmotor brachte es auf 81 kW (110 PS) bei 7.000 Umdrehungen pro Minute. Sein maximales Drehmoment von 130 Nm lag bereits bei 3.500 Umdrehungen pro Minute. 1968 wurde der Wankelmotor leistungsgesteigert und gab 94 kW (128 PS) ab.

The engine

Engine specifications

Type:	Rotary piston engine
	(Wankel)
Number of chambers:	2
Chamber volume:	2 x 491 cm³
Power:	81 kW (110 hp)
	at 7000/min

The Mazda engineers took their time to develop their own Wankel engine. This resulted in an engine with excellent running characteristics and a high reliability, which many Wankel engines previously lacked. Most failures were caused by sealing problems, which the Japanese had under control from the very beginning. The first prototype, called "L8A," had a chamber volume of 2 x 398 cm³, which was increased to 2 x 491 cm³ for the first production engine (L10A). Instead of relying on a combination of side- and peripheral intakes, the L8A and L10A only had side intakes. This resulted in a better torque and better running characteristics in the lower speed range. The twin-rotor rotary piston engine provided 81 kW (110 hp) at 7000 rpm. The maximum torque of 130 Nm was already achieved at 3500 rpm. In 1968, the Wankel engine was made more powerful and provided 94 kW (128 hp).

Der 110 S Cosmo Sport bei der Verladung in Hiroshima.
Shipping of the 110 S Cosmo Sport in Hiroshima.

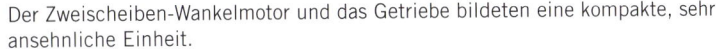

Der Zweischeiben-Wankelmotor und das Getriebe bildeten eine kompakte, sehr
ansehnliche Einheit.
The twin-rotor Wankel engine and the transmission form a compact and very
respectable unit.

Unter der Motorhaube des Mazda 110 S Cosmo Sport sah es ausgesprochen
aufgeräumt aus.
The engine compartment of the Mazda 110 S Cosmo Sport looked decidedly tidy.

Mit dem 110 S Cosmo Sport begründete Mazda seine Wankel-Tradition und
schuf gleichzeitig das erste japanische Serienautomobil mit diesem Motoren-
konzept.

With the 110 S Cosmo Sport, the Mazda company established its Wankel
engine tradition and simultaneously built the first Japanese production car
with this engine concept.

Auf den Rennstrecken feierten die NSU-Prinz-Ableger TT und TTS zahlreiche Erfolge.

The NSU Prinz derivatives TT and TTS gained numerous successes on the racing tracks.

NSU TT

Fahrzeugdaten

Hersteller:	NSU
Land:	Deutschland
Modell:	TT
Bauzeit:	1967–1972
Länge:	3.793 mm
Breite:	1.490 mm
Höhe:	1.364 mm
Radstand:	2.250 mm
Leergewicht:	685 kg
Antriebsart:	Hinterrad
Höchstgeschwindigkeit:	155 km/h
Verbrauch:	8,5 Liter/100 km

Klein, sportlich und bezahlbar – der NSU TT war eine Art Golf GTI der 1960er-Jahre. Für die junge Generation war der Zweitürer mit dem luftgekühlten Heckmotor zu jener Zeit ein Traumauto. Denn die kompakte Limousine aus Neckarsulm überzeugte mit ihrem drehfreudigen Triebwerk und bot ein bemerkenswert gutes Handling.

Seit 1963 produzierte NSU den Kleinwagen Prinz 1000, der über einen modernen Vierzylinder-Motor mit 996 cm³ Hubraum verfügte. Dieses Antriebsaggregat erwies sich schnell als sehr entwicklungsfähig. 1965 feierte mit dem 40 kW (55 PS) starken Prinz 1000 TT eine sportliche Variante des NSU seine Premiere. Das Kürzel „TT" war ein Verweis auf einstige motorsportliche Erfolge der Marke NSU bei der Tourist Trophy.

Ab 1967 firmierte der sportliche Prinz nur noch unter dem Namen „TT". Zudem verfügte er nun über 1,2 Liter Hubraum und 48 kW (65 PS) – 7 kW (10 PS) mehr als sein Vorgänger. Bis 1972 entstanden rund 50.000 Exemplare des NSU TT.

Specifications

Manufacturer:	NSU
Country:	Germany
Model:	TT
Produced:	1967–1972
Length:	3793 mm
Width:	1490 mm
Height:	1364 mm
Wheelbase:	2250 mm
Empty weight:	685 kg
Type of drive:	Rear-wheel drive
Max. speed:	155 km/h
Fuel consumption:	8.5 l/100 km

Being small, sporty and affordable, the NSU TT was something like a Golf GTI of the 1960s. The two-door car with the air-cooled rear-mounted engine was a dream car of the young generation as the compact sedan from Neckarsulm boasted a high-speed engine and offered exceptionally good handling.

Since 1963, NSU had produced the subcompact car Prinz 1000 with a modern 4-cylinder engine with a displacement of 996 cm³. This engine proved to be capable of development. In 1965, a sporty version with 40 kW (55 hp) debuted, namely the Prinz 1000 TT. The letters "TT" referred to the motorsports success of the NSU brand at the Tourist Trophy.

From 1967 on, the sporty Prinz was simply called "TT." Now it had a displacement of 1.2 liters and provided 48 kW (65 hp) – 7 kW (10 hp) more than its predecessor. Until 1972, approx. 50,000 NSU TT were built.

Der Motor

Motordaten

Bauart:	Viertakt-Reihenmotor
Zylinderzahl:	4
Hubraum:	1.177 cm³
Bohrung:	75 mm
Hub:	66,6 mm
Leistung:	48 kW (65 PS)
	bei 5.500/min

Die NSU-Ingenieure entwickelten das Triebwerk des TT aus dem luftgekühlten Reihen-Vierzylinder des Prinz 1000. Der quer im Heck eingebaute Motor des 1967 vorgestellten Sport-Modells verfügte über 1,2 Liter Hubraum und leistete 48 kW (65 PS) bei 5.500/min. Das maximale Drehmoment betrug 88 Nm, die der NSU bei 3.500 Umdrehungen lieferte. Die oben liegende Nockenwelle wurde über eine Kette angetrieben, pro Zylinder besaß der Motor zwei Ventile. Zwei Solex-Fallstromvergaser sorgten für das benötigte Benzin-Luft-Gemisch.

Der NSU wog nur 685 Kilogramm – und der TT-Motor sorgte für flotte Fahrleistungen. In 13 Sekunden beschleunigte der Wagen von 0 auf 100 km/h, sein Top-Speed betrug 155 km/h. Noch schneller war der NSU TTS, der aus 1,0 Liter Hubraum serienmäßig 51 kW (70 PS) produzierte. Der TTS war speziell für die Bedürfnisse aktiver Motorsportler entwickelt worden. NSU bot bereits ab Werk eine Version mit 62 kW (85 PS) für den Sporteinsatz an.

The engine

Engine specifications

Type:	Four-stroke in-line engine
Number of cylinders:	4
Displacement:	1177 cm³
Bore:	75 mm
Stroke:	66.6 mm
Power:	48 kW (65 hp)
	at 5500 rpm

The NSU engineers developed the engine of the TT from the air-cooled 4-cylinder in-line engine of the Prinz 1000. The engine of the sports model introduced in 1967 was mounted across in the rear, had a displacement of 1.2 liters und provided 48 kW (65 hp) at 5500 rpm. The maximum torque of 88 Nm was reached at 3500 rpm. The overhead camshaft was driven by a chain. Each cylinder had two valves. Two Solex downdraft carburetors provided the required gasoline-air mixture.

The NSU weighed only 685 kg, and the TT engine provided vivid driving characteristics. It accelerated from 0 to 100 km/h in 13 seconds, and its top speed amounted to 155 km/h. The NSU TTS was even faster. The standard version generated 51 kW (70 hp) from a displacement of 1.0 liters. As the TTS was particularly designed for the requirements of motorsports drivers, NSU also offered a 62 kW version (85 hp) ex-factory for the use in sports.

Der NSU TT war ab 1967 mit einem 48 kW (65 PS) starken 1,2-Liter-Vierzylinder ausgestattet.
From 1967 on, NSU TT was equipped with a 1.2-liter 4-cylinder engine with 48 kW (65 hp).

Der NSU TT war in den 1960er-Jahren ein Traumwagen der jungen Generation. Er war sportlich, aber bezahlbar – und damit eine Art Vorläufer des Golf GTI. Im Motorsport feierte der NSU zahlreiche Erfolge.

The NSU TT was the dream car of the young generation in the 1960s. It was sporty but affordable and hence can be seen as a predecessor of the Golf GTI. In motorsports, the NSU was also very successful.

Der Lamborghini Miura
P400 S wurde von 1968 bis
1971 gebaut. Er zählte zu
den schnellsten Sportwagen
seiner Zeit.

The Lamborghini Miura
P400 S was built from 1968
to 1971. It was among the
fastest sports cars of its time.

Lamborghini Miura S 1968

Fahrzeugdaten

Hersteller:	Lamborghini
Land:	Italien
Modell:	Miura S
Baujahr:	1968–1971
Länge:	4.370 mm
Breite:	1.760 mm
Höhe:	1.050 mm
Radstand:	2.500 mm
Leergewicht:	1.180 kg
Antriebsart:	Hinterrad
Höchstgeschwindigkeit:	274 km/h (Miura P400)
Verbrauch:	k.A.

Der Miura war das dritte Modell des exklusiven italienischen Sportwagenherstellers Lamborghini. Er wurde von 1966 bis 1972 in vier Varianten insgesamt 764-mal in hochqualitativer Handarbeit gefertigt. Mit seinem quer eingebauten V12-Mittelmotor war er eines der schnellsten Fahrzeuge seiner Zeit. Gleichzeitig war er das erste Modell, mit dem Lamborghini Gewinne erzielen konnte. Damit sicherte er den Fortbestand der Sportwagenschmiede und ermöglichte die Entwicklung weiterer Modelle.

Als der Miura 1966 sein Debüt auf dem Genfer Auto Salon feierte, begeisterten sich alsbald die Reichen und Schönen der Welt für ihn. Der Schah von Persien fuhr ebenso einen Miura wie Rod Steward und Frank Sinatra. In Deutschland war der Miura ab 1967 erhältlich. Sein empfohlener Verkaufspreis lag damals bei 75.500 DM. Heute erreichen die Modelle Verkaufspreise von bis 1,5 Millionen Euro.

Specifications

Manufacturer:	Lamborghini
Country:	Italy
Model:	Miura S
Produced:	1968–1971
Length:	4370 mm
Width:	1760 mm
Height:	1050 mm
Wheelbase:	2500 mm
Empty weight:	1180 kg
Type of drive:	Rear-wheel drive
Max. speed:	274 km/h (Miura P400)
Fuel consumption:	n.a.

The Miura was the third model of the exclusive Italian sports car manufacturer Lamborghini. From 1966 to 1972, four versions with a total of 764 units were handcrafted with high quality. The V12 midship engine mounted at right angle made it one of the fastest vehicles of its time. It was also the first model that yielded profit for Lamborghini. This guaranteed the further existence of the sports car forge and made it possible to develop more models.

When the Miura debuted at the Geneva International Motor Show in 1966, the rich and beautiful began soon to fancy this car. The Shah of Persia, Rod Steward and Frank Sinatra, they all rode a Miura. In Germany, the Miura was first available in 1967. The suggested retail price amounted to 75,500 DM. Today, these models are sold for prices up to 1.5 million Euros.

Der Motor

Motordaten

Bauart:	Viertakt-Otto-V-Motor
Zylinderzahl:	12
Hubraum:	3.929 cm³
Bohrung:	82 mm
Hub:	62 mm
Leistung:	272 kW (370 PS)
	bei 7.700/min

In allen Miura-Modellen kam ein V12-Motor mit einem Gabelwinkel von 60 Grad zum Einsatz. Der hinter den Sitzen quer eingebaute Mittelmotor sorgte für eine hervorragende Gewichtsverteilung, die sich positiv in den Fahrleistungen niederschlug, sowie für einen entsprechenden Geräuschpegel in der Fahrgastzelle. Der wassergekühlte Motor des bis 1969 gefertigten P400 besaß zwei oben liegende kettengetriebene Nockenwellen pro Zylinderreihe und übertrug bei 7.000 Umdrehungen pro Minute 260 kW (350 PS) über ein vollsynchronisiertes Fünfganggetriebe auf die Hinterachse. Der Wagen beschleunigte innerhalb von 6,7 Sekunden auf 100 km/h.

Der Motor des von 1968 bis 1971 ausgelieferten Miura P400 S wurde auf 272 kW (370 PS) bei 7.700 Umdrehungen pro Minute gesteigert. Im von 1971 bis 1972 gebauten P400 SV versah eine verbesserte Version ihren Dienst. Sie brachte es auf 283 kW (385 PS) bei 7.700 Umdrehungen pro Minute.

The engine

Engine specifications

Type:	4-stroke Otto V-engine
Number of cylinders:	12
Displacement:	3929 cm³
Bore:	82 mm
Stroke:	62 mm
Power:	272 kW (370 hp)
	at 7700/min

All Miura models had a V12 engine with an angle of 60°. The midship engine was mounted at right angle behind the seats and thus provided for an excellent weight balance, which had a positive influence on the driving characteristics but also resulted in a high level of noise in the passenger cell. The water-cooled engine of the P400 built until 1969 had two chain-driven overhead camshafts per cylinder bank. At 7000 rpm, it transferred 260 kW (350 hp) to the rear axle via a fully synchronized five-gear transmission. The car accelerated to 100 km/h in 6.7 seconds.

In the Miura P400 S shipped from 1968 to 1971, the engine power was increased to 272 kW (370 hp) at 7700 rpm. The P400 SV built from 1971 to 1972 had an even more improved engine that provided 283 kW (385 hp) at 7700 rpm.

Der V12-Motor des Lamborghini P400 SV brachte es auf 283 kW (385 PS).
The V12 engine of the Lamborghini P400 SV provided 283 kW (385 hp).

Volvo P1800 E und P1800 ES.
Volvo P1800 E and P1800 ES.

Volvo P1800 ES „Schneewittchensarg"   1971

Volvo P1800 ES „Schneewittchensarg" 1971

Fahrzeugdaten

Hersteller:	Volvo
Land:	Schweden
Modell:	P1800 ES
Baujahr:	1971–1973
Länge:	4.390 mm
Breite:	1.700 mm
Höhe:	1.315 mm
Radstand:	2.450 mm
Leergewicht:	1.190 kg
Antriebsart:	Hinterrad
Höchstgeschwindigkeit:	ca. 190 km/h
Verbrauch:	11–14 Liter/100 km

Der Volvo P1800 ES ist im Grunde genommen eine Modellvariante des seit 1961 gebauten P1800-Sportcoupes. Für die Entwicklung der Heckpartie des P1800 ES zeichnete der Designer Jan Wilsgaard verantwortlich, der das Fahrzeug mit seiner vollständig aus Glas gefertigten Heckklappe zu einer einzigartigen Erscheinung machte. Ihr hatte der sportliche Kombi auch seinen von einer deutschen Automobilzeitschrift verliehenen Spitznamen „Schneewittchensarg" zu verdanken. Die wenig schmeichelhafte Bezeichnung war womöglich der Grund, warum der P1800 ES in Deutschland nur wenige Käufer fand. Umso mehr konnten sich die Amerikaner für ihn begeistern. Als 1973 bekannt wurde, dass seine Produktion wegen der Einführung neuer amerikanischer Sicherheitsbestimmungen eingestellt werden würde, waren die noch vorhandenen fabriksneuen Modelle binnen weniger Wochen ausverkauft. Vom Volvo P1800 ES wurden von 1971 bis 1973 8.077 Stück gebaut.

Specifications

Manufacturer:	Volvo
Country:	Sweden
Model:	P1800 ES
Produced:	1971–1973
Length:	4390 mm
Width:	1700 mm
Height:	1315 mm
Wheelbase:	2450 mm
Empty weight:	1190 kg
Type of drive:	Rear-wheel drive
Max. speed:	approx. 190 km/h
Fuel consumption:	11–14 liters/100 km

Basically, the Volvo P1800 ES is a variation of the sport coupe P1800 built since 1961. The designer responsible for the development of the rear part of the P1800 ES was Jan Wilsgaard, who came up with the rear door made completely of glass. This gave the car its unique appearance and caused a German car magazine to dub the vehicle "Snow White's coffin." This not exactly flattering nickname might have been the reason why the P1800 ES did not attract many customers in Germany. The Americans, on the other hand, loved this car. When it transpired in 1973 that the production was discontinued due to the introduction of new American security regulations, the existing new units were completely sold within a few weeks. From 1971 to 1973, 8077 units of the Volvo P1800 ES were built.

Der Motor

Motordaten

Bauart:	Viertakt-Otto-Reihenmotor
Zylinderzahl:	4
Hubraum:	1.986 cm³
Bohrung:	88,9 mm
Hub:	80 mm
Leistung:	91 kW (124 PS)
	bei 6.000/min

In der europäischen Version des Volvo P1800 ES kam das Aggregat B20 E zum Einsatz. Es war ein Vierzylinder-Otto-Reihenmotor mit einem Hubraum von 2 Litern. Der Motor leistete bei 6.000 Umdrehungen pro Minute 91 kW (124 PS). Sein maximales Drehmoment von 173 Nm entfaltete die Maschine bei 3.500 Umdrehungen pro Minute. Das Verdichtungsverhältnis lag bei 10,5:1. Der wassergekühlte Motor besaß eine elektronische Benzineinspritzung, die Ventilsteuerung erfolgte über eine seitliche, über Zahnräder angetriebene Nockenwelle. Mit dem B20-E-Motor wurde eine Höchstgeschwindigkeit von rund 190 km/h erreicht.

In den für die USA bestimmten P1800-ES-Einheiten wurde die Motortype B20 F verbaut. Sie zeichnete sich durch eine etwas schwächere Leistung von 85 kW (115 PS) bei 6.000 Umdrehungen pro Minute und einem etwas kleineren Drehmoment von 163 Nm bei 3.500 Umdrehungen pro Minute aus. Auch das Verdichtungsverhältnis war mit 8,7:1 etwas geringer.

The engine

Engine specifications

Type:	4-stroke in-line Otto engine
Number of cylinders:	4
Displacement:	1986 cm³
Bore:	88.9 mm
Stroke:	80 mm
Power:	91 kW (124 hp)
	at 6000/min

In the European version of the Volvo P1800 ES, the B20 E engine was used. This was a 4-cylinder in-line Otto engine with a displacement of 2 liters. At 6000 rpm, it provided 91 kW (124 hp). Its maximum torque of 173 Nm was achieved at 3500 rpm. The compression ratio amounted to 10.5:1. The water-cooled engine used electronic fuel injection. Valve control was carried out by a gear-driven camshaft mounted at the side. With the B20 E engine, the car achieved a maximum speed of approx. 190 km/h.

In the P1800 ES models for the US market, a B20 F engine was used. With 85 kW (115 hp) at 6000 rpm and 163 Nm at 3500 rpm, its power and torque were both a little smaller than that of the European version, as well as the compression ration of only 8.7:1.

Der Reihenmotor des Volvo P1800 ES leistete
in seiner Europa-Variante 91 kW.

The European version of the in-line engine of
the Volvo P1800 ES provided 91 kW.

Black
magic

Zeitgeist: Mit dieser Anzeige warb Opel 1975 für das Manta-Sondermodell „Black Magic".

In 1975, Opel marketed the "Black Magic" special version of the Manta with ads like this.

Das Topmodell der Manta-Baureihe war der GT/E. Sein 1,9-Liter-Motor leistete 77 kW (105 PS).

The top model of the Manta series was the version GT/E. Its 1.9-liter engine provided 77 kW (105 hp).

Opel Manta

Fahrzeugdaten

Hersteller:	Opel
Land:	Deutschland
Modell:	Manta A
Bauzeit:	1970–1975
Länge:	4.292 mm
Breite:	1.626 mm
Höhe:	1.355 mm
Radstand:	2.430 mm
Leergewicht:	950 kg
Antriebsart:	Hinterrad
Höchstgeschwindigkeit:	164 km/h
Verbrauch:	8,6 Liter/100 km

*Werte für 1972 Opel Manta A 1,6 S

In den USA feierte ab Mitte der 1960er-Jahre eine neue Fahrzeugklasse große Erfolge: die Pony Cars. Diese Bezeichnung stand für Autos, die eindeutig auf den Geschmack der wachsenden Zahl junger und zahlungskräftiger Käufer ausgerichtet waren. Die neuen Modelle verbanden sportliche Optik mit bewährter Großserientechnik.

Auch in Deutschland schien es für dieses Konzept einen Markt zu geben, denn ab 1968 bot Ford mit dem Capri eine europäische Variante des Pony Cars an. Opel wollte der Konkurrenz das neue Feld nicht ohne Gegenwehr überlassen. Und so stellten die Rüsselsheimer 1970 den Manta vor. Das Coupé basierte technisch auf der ebenfalls neuen Limousine Ascona, sollte aber von den Kunden als eigenständiges Modell wahrgenommen werden. Unter der gefälligen Karosserie fand sich Technik aus dem Opel-Baukasten. Diese Mischung kam beim Publikum an: Bis 1975 verkaufte Opel fast eine halbe Million Exemplare des Coupés, bevor der Manta B die Nachfolge antrat.

Specifications

Manufacturer:	Opel
Country:	Germany
Model:	Manta A
Produced:	1970–1975
Length:	4292 mm
Width:	1626 mm
Height:	1355 mm
Wheelbase:	2430 mm
Empty weight:	950 kg
Type of drive:	Rear-wheel drive
Max. speed:	164 km/h
Fuel consumption:	8.6 l/100 km

* data for the 1972 Opel Manta A 1.6 S

Since the mid-1960s, the new vehicle class called "pony cars" was highly successful in the USA. The designation referred to cars that were tailored to the taste of a growing number of young and solvent buyers. The new models combined a sportive appearance with proven mass production technology.

Obviously, there was a market for this concept in Germany as well. Beginning with 1968, Ford sold the Capri as a European type of pony car. Opel did not want to leave the field clear for the competitors without trying to fight back, and thus in 1970, the company from Rüsselsheim introduced the Manta. The technology of the coupé was based on the new Ascona sedan but Opel wanted the customers to view the Manta as an independent model in its own right. Below the pleasant auto body, technology from the Opel modular construction kit was used.

This mix was well received by the audience: Opel nearly sold half a million units of the coupé until 1975, when the successor model Manta B was introduced.

Der Motor

Motordaten

Bauart:	Viertakt-Reihenmotor
Zylinderzahl:	4
Hubraum:	1.584 cm³
Bohrung:	85 mm
Hub:	69,8 mm
Leistung:	59 kW (80 PS)
	bei 5.200/min

Als Opel den neuen Manta im September 1970 in Timmendorfer Strand der Presse vorstellte, boten die Rüsselsheimer das Coupé mit drei Motorisierungen an: Das Basismodell besaß den 1,6-Liter-Vierzylinder mit 50 kW (68 PS), während die SR-Variante bei gleichem Hubraum immerhin 59 kW (80 PS) leistete. Die Topversion verfügte über den 1,9-Liter-Reihenvierzylinder aus dem Opel Rekord, der 66 kW (90 PS) bei 5.100/min produzierte. Motorblock und Zylinderkopf der Manta-Triebwerke waren jeweils aus Grauguss gefertigt. Ab 1972 kam ein Einstiegsmodell mit 1,2-Liter-Hubraum und 44 kW (60 PS) dazu. Im Herbst 1973 präsentierte Opel den sportlichsten Manta mit Namen GT/E. Das 1,9-Liter-Vierzylinder-Triebwerk verfügte über eine elektronische Benzineinspritzung, die dem Opel zu 77 kW (105 PS) verhalf. Das 980 Kilogramm schwere Coupé erreichte respektable Fahrleistungen: Von 0 auf 100 km/h benötigte der GT/E 10,5 Sekunden, seine Höchstgeschwindigkeit betrug 188 km/h.

The engine

Engine specifications

Type:	Four-stroke in-line engine
Number of cylinders:	4
Displacement:	1584 cm³
Bore:	85 mm
Stroke:	69.8 mm
Power:	59 kW (80 hp)
	at 5200 rpm

When Opel presented the new Manta to the press in Timmendorfer Strand in September 1970, the coupé was available with three different engines. The base model had a 1.6-liter 4-cylinder engine with 50 kW (68 hp), while the SR version used the same displacement to provide 59 kW (80 hp). The top version was equipped with a 1.9-liter 4-cylinder in-line engine from the Opel Rekord, which generated 66 kW (90 hp) at 5100 rpm. Engine block and cylinder heads of the Manta engines were made of grey cast iron. Beginning with 1972, an entry-level model with a displacement of 1.2 liters and 44 kW (60 hp) supplemented the portfolio. In autumn 1973, Opel presented the sportiest Manta, namely the GT/E. Its 1.9-liter 4-cylinder engine had an electronic fuel injection and generated 77 kW (105 hp). The coupé of 980 kg achieved a respectable driving performance. The GT/E needed 10.5 seconds to accelerate from 0 to 100 km/h, and its top speed amounted to 188 km/h.

Ab Beginn der Serienproduktion im Jahr 1970 bot Opel den Manta mit dem bewährten 1,6-Liter-Reihen-Vierzylinder an.
Beginning with series production in 1970, Opel sold the Manta with the proven 1.6-liter 4-cylinder in-line engine.

Beim Manta packte Opel 1970 bewährte Technik in eine attraktive Hülle. Das Coupé sollte dem Ford Capri Paroli bieten – und das gelang. Bis 1975 baute Opel fast eine halbe Million Exemplare der Manta-Baureihe.

For the Manta, Opel wrapped proven technology in an attractive shell. The coupé was pitted against the Ford Capri, and successfully so. Until 1975, Opel built nearly half a million Manta.

Als erstes deutsches Serien-Auto war der BMW 2002 Turbo 1973 mit einem Abgasturbolader ausgerüstet.

In 1973, the BMW 2002 Turbo was the first German standard car with an exhaust turbo charger.

BMW 2002 Turbo

1973

Fahrzeugdaten

Hersteller:	BMW
Land:	Deutschland
Modell:	2002 Turbo
Bauzeit:	1973–1974
Länge:	4.220 mm
Breite:	1.620 mm
Höhe:	1.410 mm
Radstand:	2.500 mm
Leergewicht:	1.080 kg
Antriebsart:	Hinterrad
Höchstgeschwindigkeit:	211 km/h
Verbrauch:	14,7 Liter/100 km

Herbst 1973: Die erste große Ölkrise traf die westlichen Industrieländer mit voller Wucht. In Deutschland verordnete die Bundesregierung unter Willy Brandt vier autofreie Sonntage mit Fahrverboten, und für ein halbes Jahr galt auf den Autobahnen Tempo 100. Genau in diese Zeit fiel die Vorstellung des neuen BMW 2002 Turbo. Als erster deutscher Serienwagen verfügte die kompakte Limousine über einen Abgasturbolader. Mit seiner Hilfe leistete der Zwei-Liter-Vierzylinder-Motor sportive 125 kW (170 PS). Für den Sprint von 0 auf 100 km/h benötigte der BMW nur knapp sieben Sekunden und erreichte eine beachtliche Höchstgeschwindigkeit von 211 km/h. Das waren beeindruckende Fahrleistungen, doch mit einem durchschnittlichen Verbrauch von rund 15 Litern auf 100 Kilometer passte der Turbo eigentlich nicht mehr in die neue Zeit des Benzinsparens. Trotzdem wurden bis zur Einstellung der Produktion Ende 1974 letztlich immerhin 1.672 Einheiten des BMW 2002 Turbo gebaut.

Specifications

Manufacturer:	BMW
Country:	Germany
Model:	2002 Turbo
Produced:	1973–1974
Length:	4220 mm
Width:	1620 mm
Height:	1410 mm
Wheelbase:	2500 mm
Empty weight:	1080 kg
Type of drive:	Rear-wheel drive
Max. speed:	211 km/h
Fuel consumption:	14.7 l/100 km

Autumn 1973: The first oil crisis hit the western industrialized countries heavily. The German federal government led by Willy Brandt decreed four car-free Sundays with a ban on driving, and for half a year, there was a speed limit of 100 km/h on the German autobahn. At the same time, the new BMW 2002 Turbo was introduced. The compact sedan was the first German standard car with an exhaust turbo charger, by which the 2-liter 4-cylinder engine achieved a sporty power of 125 kW (170 hp). Thus, the BMW only needed seven seconds to accelerate from 0 to 100 km/h. It also had a top speed of 211 km/h. These figures were quite impressive, but with an average fuel consumption of 15 l/100 km, the Turbo did not fit the new era of fuel conservation. Nonetheless 1672 units of the BMW 2002 Turbo were built until production folded in the end of 1974.

Der Motor

Motordaten

Bauart:	Viertakt-Reihenmotor mit Turbolader
Zylinderzahl:	4
Hubraum:	1.991 cm³
Bohrung:	89 mm
Hub:	80 mm
Leistung:	125 kW (170 PS) bei 5.800/min

Leistungssteigerung mittels Abgasturbolader war bereits 1973 keine ganz neue Idee – zumindest auf internationalen Rennstrecken. Doch im deutschen Serienautomobilbau betrat BMW mit dem Motor des 2002 Turbo als Pionier Neuland – noch bevor Porsche im Jahr 1974 eine aufgeladene Variante des 911 vorstellte. Bei der Entwicklung des 2002 Turbo konnten die BMW-Techniker auf Rennerfahrungen vertrauen: Bereits 1969 hatten die Münchner einen 2002 mit Turbomotor in der Tourenwagen-Europameisterschaft eingesetzt. Der 205 kW (280 PS) starke Vierzylinder-Zwei-Liter-Turbo bescherte dem österreichischen BMW-Werksfahrer Dieter Quester den EM-Titel. Deutlich ziviler präsentierte sich die Serienausführung des BMW 2002 Turbo, die auf der IAA in Frankfurt am Main 1973 ihre Premiere feierte. Der Reihen-Vierzylinder brachte es auf 125 kW (170 PS), die bei 5.800 Umdrehungen pro Minute abgegeben wurden. Sein maximales Drehmoment von 240 Nm erreichte das bayerische Kraftpaket bei 4.000/min.

The engine

Engine specifications

Type:	Four-stroke in-line engine with turbo charger
Number of cylinders:	4
Displacement:	1991 cm³
Bore:	89 mm
Stroke:	80 mm
Power:	125 kW (170 hp) at 5.800/min

Increasing the power by means of an exhaust turbo charger was not a really new idea in 1973, at least in international racing. In the area of German standard cars, however, BMW did pioneering work with the engine of the 2002 Turbo. This was even before Porsche introduced a charged variant of the 911 in 1974. In developing the 2002 Turbo, the BMW engineers could rely on their racing experiences: the company in Munich had already deployed a 2002 with turbo engine in the European touring car championship of 1969. The 4-cylinder 2-liter turbo engine with 205 kW (280 hp) helped the Austrian BWM company driver Dieter Quester to win the European championship. The standard version of the BMW 2002 Turbo, which was introduced at the IAA in Frankfurt/Main in 1973, presented itself in a much more innocent way. The 4-cylinder in-line engine produced 125 kW (170 hp) at 5800 rpm. The Bavarian power pack achieved its maximum torque of 240 Nm at 4000 rpm.

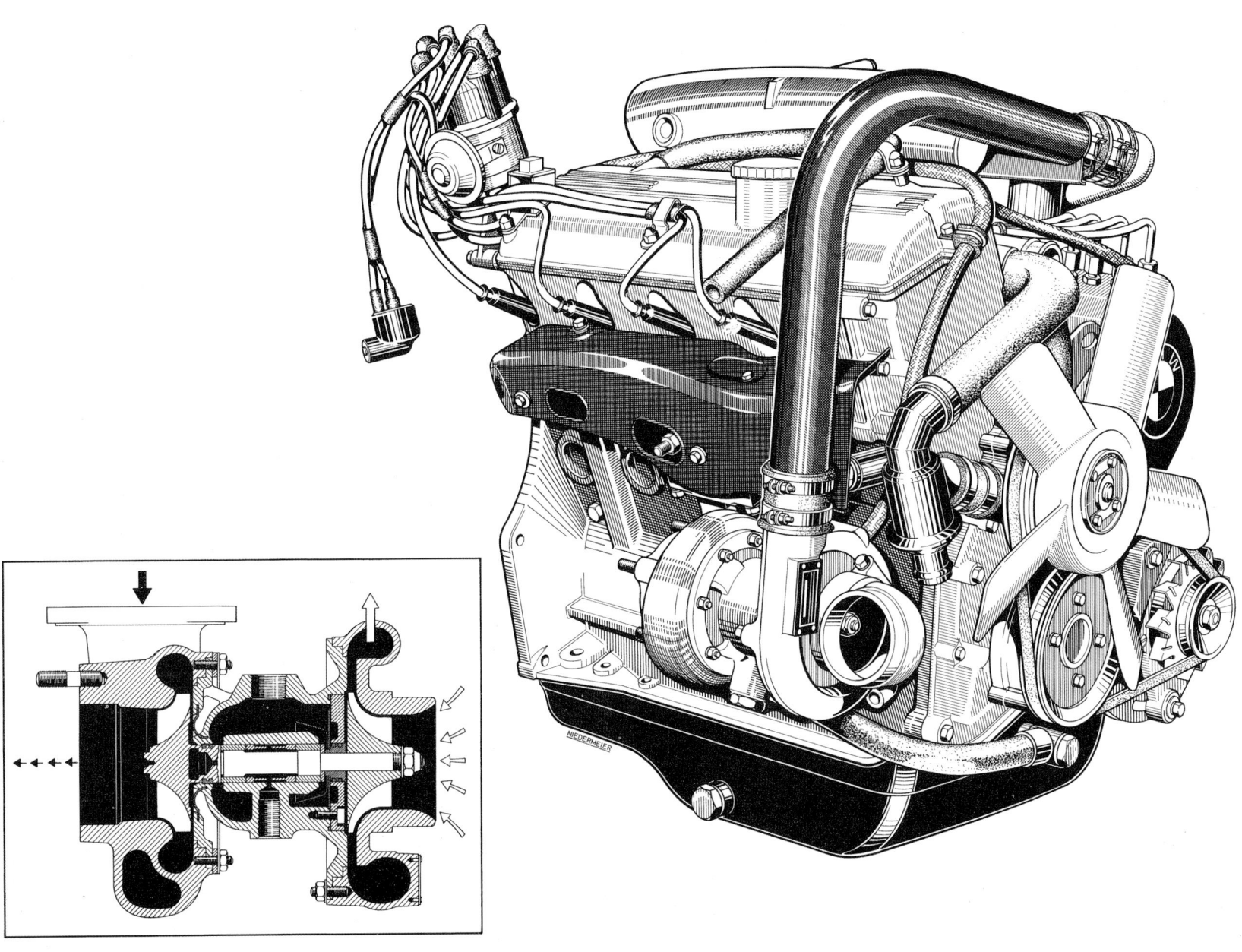

Der 2,0-Liter-Vierzylinder der Sport-Limousine leistete mithilfe des Abgasturboladers 125 kW (170 PS).

By means of an exhaust turbo charger, the 2.0-liter 4-cylinder sports sedan provided 125 kW (170 hp).

BMW testete die Turbo-Technologie zunächst im Motorsport. Mit einem aufgeladenen 2002 feierten die Münchner 1969 große Rennerfolge in der Tourenwagen-Europameisterschaft. Vier Jahre später ging der 2002 Turbo in Serie.

BMW first tested the turbo technology in motorsports. With a turbocharged 2002, the Munich company enjoyed a huge racing success in the European touring car championship in 1969. Four years later, the 2002 Turbo went into series production.

Volvo 240 von 1983 in der Ausführung als Limousine.
Sedan version of the 1983 Volvo 240.

Volvo 240 1974

Fahrzeugdaten

Hersteller:	Volvo
Land:	Schweden
Modell:	240
Baujahr:	1974–1993
Länge:	4.790 mm (Kombi)
Breite:	1.710 mm
Höhe:	1.460 mm
Radstand:	2.640 mm
Leergewicht:	1.440 kg
Antriebsart:	Hinterrad
Höchstgeschwindigkeit:	190 km/h
Verbrauch:	ca. 8,5 Liter/100 km

Der Volvo 240 war ab 1974 als Limousine und Kombi erhältlich. Während seiner 19-jährigen Bauzeit wurde die Mittelklassereihe, zu der die Typen 242, 244 und 245 gehörten, laufend weiterentwickelt. Die 240er-Reihe war in Europa und den USA äußerst beliebt. Sie verkörperte auch das, was man unter einem sehr sicheren Auto verstand. So kürte etwa die US-Verkehrssicherheitsbehörde den Volvo 240 im Jahr 1976 als Referenz für die Festlegung künftiger Personenwagen-Sicherheitsstandards.

Den Skandinaviern war es mit dem 240er gelungen, der Marke Volvo einen unverkennbaren Stempel aufzudrücken. Die erstmals dreigeteilte Frontpartie mit der schräg verlaufenden Zierleiste über dem Kühlergrill ist im Prinzip bis heute erhalten geblieben. Sie geben den Attributen, Sicherheit, Qualität und Komfort ein weithin sichtbares Kennzeichen.

Durch sein funktionelles Design hob sich der 240er von seinen Mitbewerbern ab, und es machte ihn zu einem Fahrzeug für Individualisten.

Specifications

Manufacturer:	Volvo
Country:	Sweden
Model:	240
Produced:	1974–1993
Length:	4790 mm (station wagon)
Width:	1710 mm
Height:	1460 mm
Wheelbase:	2640 mm
Empty weight:	1440 kg
Type of drive:	Rear-wheel drive
Max. speed:	190 km/h
Fuel consumption:	approx. 8.5 liters/100 km

Beginning with 1974, the Volvo 240 was available as sedan and as station wagon. During its 19 years of production, the middle class series comprising the types 242, 244 und 245 was permanently advanced. The 240 series was extremely popular in Europe and the USA. It was the epitome of a very safe car. In 1976, the US traffic safety administration selected the Volvo 240 as reference model to define future safety standards for passenger cars.

With the 240, the Scandinavian manufacturer succeeded in leaving a distinctive mark on the Volvo brand. The three-part front with the oblique trim strip across the radiator grille debuted with this model and is basically still in use today. This design has become highly visible characteristic signaling safety, quality and comfort.

The functional design clearly distinguished the 240 from its competitors and made it a car for individualists.

Der Motor

Motordaten

Bauart:	Otto-Turbo-Reihenmotor
Zylinderzahl:	4
Hubraum:	2.127 cm³
Bohrung:	92 mm
Hub:	80 mm
Leistung:	114 kW (155 PS)
	bei 5.500/min

Für die Modellreihe 240 bot Volvo Otto- und Dieselmotoren verschiedener Leistungsklassen an. Während der Produktionszeit wurde das Motorenprogramm laufend modernisiert. Allen Aggregaten gemeinsam war die hohe Lebensdauer. Bei den Vierzylinder-Ottomotoren galt eine Laufleistung von 300.000 Kilometern als normal, selbst mehr als 500.000 Kilometer waren nicht außergewöhnlich.

In den ersten 240ern wurden noch die Ottomotoren der 144er-Vorgängerserie eingebaut. Sie brachten es auf 60 kW (82 PS). Ab 1980 war für den 240 mit dem B21ET der leistungsstärkste Ottomotor verfügbar. Dank seines Turbos brachte er es auf stolze 114 kW (155 PS), die er bei 5.500 Umdrehungen pro Minute freigab. Das Aggregat war auch in Sachen Drehmoment ein Kraftpaket: Es lag bei 3.750 Umdrehungen pro Minute bei 240 Nm. Der wassergekühlte Motor besaß eine oben liegende Nockenwelle, über die je zwei Ventile pro Zylinder gesteuert wurden. Das Verdichtungsverhältnis der Maschine lag bei 7,5:1.

The engine

Engine specifications

Type:	In-line Otto turbo engine
Number of cylinders:	4
Displacement:	2127 cm³
Bore:	92 mm
Stroke:	80 mm
Power:	114 kW (155 hp)
	at 5500/min

With the 240 series, Volvo offered Otto and Diesel engines of various classes. During production time, the engine program was permanently modernized. A common attribute of all engines was the long life. A mileage of 300,000 km was deemed normal for the 4-cylinder Otto engines; even 500,000 km was not considered extraordinary.

The first 240s still used Otto engines of the previous 144 series with 60 kW (82 hp). Beginning in 1980, the more powerful B21ET Otto engine was available for the 240. Due to its turbocharger, the engine provided a whopping 114 kW (155 hp) at 5500 rpm. With 240 Nm at 3750 rpm, the torque was also impressive. The water-cooled engine had an overhead camshaft controlling the two valves per cylinder. The compression ratio amounted to 7.5:1.

Der 114-kW-Otto-Turbo-Reihenmotor B21ET wurde für die Modellreihe 240 ab 1980 angeboten.
The 114 kW in-line Otto turbo engine B21ET was available for the 240 series beginning with 1980.

Mercedes Benz 450 SEL 6.9 in der Version von 1980.
A Mercedes Benz 450 SEL 6.9 of the 1980 version.

Mercedes-Benz 450 SEL 6.9 1975

Fahrzeugdaten

Hersteller:	Mercedes-Benz
Land:	Deutschland
Modell:	450 SEL 6.9
Baujahr:	1975–1980
Länge:	5.060 mm
Breite:	1.870 mm
Höhe:	1.410 mm
Radstand:	2.960 mm
Leergewicht:	1.935 kg
Antriebsart:	Hinterrad
Höchstgeschwindigkeit:	225 km/h
Verbrauch:	rund 20 Liter/100 km

Im Herbst 1972 schlug mit dem W116 die Geburtsstunde der neuen Mercedes-Fahrzeuggeneration der Oberklasse – der S-Klasse. Mit der Modellreihe W116 verabschiedete sich Mercedes vom konservativen Design seiner Modelle. Ihre Formgebung war zukunftsweisend, und im Hinblick auf die passive Sicherheitstechnik waren die neuen Mercedes-Fahrzeuge ein Meilenstein. Das Flaggschiff der W116-Reihe, der 450 SEL 6.9, wurde im Frühjahr 1973 der Öffentlichkeit vorgestellt. Es war seinerzeit die schnellste Limousine der Welt. Wegen der kurz darauf eintretenden ersten Ölkrise sah sich Mercedes jedoch veranlasst, das Spitzenmodell erst 1975 auf den Markt zu bringen.

1975 betrug der Listenpreis für die Basisversion des 450 SEL 6.9 69.930 DM. Der Wagen überzeugte durch seinen hohen Fahrkomfort, der durch die Hydropneumatik-Federung erreicht wurde. Von der Reihe W116 wurden bis 1980 mehr als 473.000 Stück gebaut. Auf den 450 SEL 6.9 entfielen annähernd 7.400 Einheiten.

Specifications

Manufacturer:	Mercedes-Benz
Country:	Germany
Model:	450 SEL 6.9
Produced:	1975–1980
Length:	5060 mm
Width:	1870 mm
Height:	1410 mm
Wheelbase:	2960 mm
Empty weight:	1935 kg
Type of drive:	Rear-wheel drive
Max. speed:	225 km/h
Fuel consumption:	approx. 20 liters/100 km

In autumn 1972, the W116 heralded the new generation of Mercedes luxury vehicles, namely the S class. With this series, Mercedes abandoned its previous conservative design. The shape was forward-looking. Furthermore, the new Mercedes cars represented a milestone with respect to passive safety technology. The flagship of the W116 series was the 450 SEL 6.0, which was introduced in public in spring 1973. At this time, it was the fastest sedan in the world. However, due to the first oil crisis that took place soon afterwards, Mercedes was forced to launch the top model not before 1975.

In 1975, the list price of the standard version amounted to 69,930 DM. The hydropneumatic suspension provided for a highly comfortable ride. Until 1980, more then 473,000 units of the W116 series were built. Among these were approx. 7400 units of the 450 SEL 6.9.

Der Motor

Motordaten

Bauart:	Viertakt-V8-Ottomotor
Zylinderzahl:	8
Hubraum:	6.834 cm³
Bohrung:	107 mm
Hub:	95 mm
Leistung:	210 kW (286 PS)
	bei 3.000/min

Dem Mercedes 450 SEL 6.9 wurde ein mächtiger V8-Ottomotor mit einem Gabelwinkel von 90 Grad spendiert. Sein Hubraum lag bei 6.834 cm³, seine Verdichtung bei 8,8:1. Bei einer Drehzahl von 3.000 Umdrehungen pro Minute gab das Aggregat seine höchste Leistung von 210 kW (286 PS) ab.

Der wassergekühlte Motor trug die Typenbezeichnung „M100". Er besaß pro Zylinder zwei in einer Reihe angeordnete Ventile. Pro Zylinderbank war eine oben angeordnete Nockenwelle vorgesehen, die über Ketten angetrieben wurde. Der Motor besaß eine Bosch-K-Jetronic-Benzineinspritzung und war mit einer Trockensumpfschmierung ausgestattet. Das Aggregat gab seine Kraft über eine Vierstufen-Automatik an die Hinterräder ab. Die Höchstgeschwindigkeit des 450 SEL 6.9 war ab Werk mit 225 km/h angegeben. Bei unabhängigen Tests wurden jedoch 237 km/h ermittelt. Die Beschleunigung vom Stillstand auf 100 km/h erfolgte binnen 7,8 Sekunden. 200 km/h wurden nach rund 34 Sekunden erreicht.

The engine

Engine specifications

Type:	4-stroke V8 Otto engine
Number of cylinders:	8
Displacement:	6834 cm³
Bore:	107 mm
Stroke:	95 mm
Power:	210 kW (286 hp)
	at 3000 rpm

The Mercedes 450 SEL 6.9 was equipped with a powerful V8 Otto engine with a cylinder bank angle of 90°, a displacement of 6834 cm³ and a compression ratio of 8.8:1. At 3000 rpm, the engine provided its maximum power of 210 kW (286 hp).

The type designation of this water-cooled engine was M100. It had two in-line valves per cylinder and one chain-driven overhead camshaft per cylinder bank. Furthermore, it was equipped with K-Jetronic fuel injection system by Bosch and dry sump lubrication. The power was transferred to the rear wheels by a four-stage automatic transmission. The maximum speed of the 450 SEL 6.9 was rated at 225 km/h by the manufacturer; in independent tests, however, a speed of up to 237 km/h was achieved. Accelerating the car to 100 km/h took 7.8 seconds. After further 34 seconds, a speed of 200 km/h could be reached.

Antrieb für die damals schnellste
Limousine der Welt: der V-8-Motor des
Mercedes-Benz 450 SEL 6.9
aus dem Jahr 1975.

The engine of the fastest sedan car of its
time: the V8 engine of the Mercedes-Benz
450 SEL 6.9 of the year 1975.

Der VW Golf GTI war mit seinen 81 kW (110 PS) in den 1970er-Jahren ein regelrechter „Volks-Sportwagen".

With its 81 kW (110 hp), the Golf GTI was a real "people's sports car" of the 1970s.

VW Golf I GTI

Fahrzeugdaten

Hersteller:	VW
Land:	Deutschland
Modell:	Golf GTI
Bauzeit:	1976–1983
Länge:	3.725 mm
Breite:	1.630 mm
Höhe:	1.395 mm
Radstand:	2.400 mm
Leergewicht:	810 kg
Antriebsart:	Vorderrad
Höchstgeschwindigkeit:	182 km/h
Verbrauch:	10,5 Liter/100 km

*Werte für 1976 VW Golf GTI

Der VW Golf stand ab 1974 für das Konzept der neuen Kompaktklasse im Automobilbau. Designer Giorgio Giugiaro gab dem Käfer-Nachfolger eine nüchterne, aber moderne Form mit kurzen Karosserie-Überhängen. Und die Wolfsburger Ingenieure sorgten für zeitgemäße Technik: wassergekühlter Frontmotor, Frontantrieb und ein entwicklungsfähiges Fahrwerk.

Für einige VW-Techniker war schnell klar, dass der Golf durchaus auch das Zeug zur sportlichen Limousine hatte – und sie begannen ohne offiziellen Auftrag mit den ersten Arbeiten an einem Sport-Golf. Mit dem 1,6-Liter-Vierzylinder aus dem Audi 80 GTE gab es im Konzern einen passenden Motor mit kräftigen 81 kW (110 PS). Und tatsächlich wurde aus dem Geheimprojekt schließlich der Golf GTI.

Ab 1976 war die sportliche Variante des Wolfsburger Erfolgsmodells lieferbar. Und der Verkaufserfolg übertraf alle Erwartungen der Macher: Bis 1983 produzierte VW von der ersten Generation des Golf GTI insgesamt 270.000 Exemplare.

Specifications

Manufacturer:	VW
Country:	Germany
Model:	Golf GTI
Produced:	1976–1983
Length:	3725 mm
Width:	1630 mm
Height:	1395 mm
Wheelbase:	2400 mm
Empty weight:	810 kg
Type of drive:	Front-wheel drive
Max. speed:	182 km/h
Fuel consumption:	10.5 l/100 km

* data for the 1976 VW Golf GTI

From 1974 on, the VW Golf epitomized the new compact car class. Its designer Giorgio Giugiaro gave the Beetle successor a sober but modern shape with short body overhangs. The engineers from Wolfsburg also provided up-to-date technology: water-cooled engine, front-wheel drive and a developable undercarriage.

Some VW technicians soon realized that the Golf also was sports car material. Without being officially commissioned, they began working on a sports Golf. The company had a suitable engine, namely the 1.6-liter 4-cylinder engine of the Audio 80 GTE with 81 kW (110 hp). Eventually, the secret project became the Golf GTI.

From 1976 on, the sports version of the success model from Wolfsburg was available. The sales figures exceeded all expectations of the manufacturer. Until 1983, VW build 270,000 units of the first generation Golf GTI.

Der Motor

Motordaten

Bauart:	Viertakt-Reihenmotor
Zylinderzahl:	4
Hubraum:	1.588 cm³
Bohrung:	79,5 mm
Hub:	80 mm
Leistung:	81 kW (110 PS)
	bei 6.100/min

Das Herzstück des GTI kam eigentlich aus Ingolstadt. Dort hatten die Motorentechniker den 1,6-Liter-Vierzylinder-Motor des Audi 80 GT mit einer Bosch-K-Jetronic ausgestattet. Mit dieser Benzineinspritzung leistete das Triebwerk 81 kW (110 PS) bei 6.100 Umdrehungen. Damit war die Audi-Entwicklung das geeignete Aggregat, um auch dem VW Golf zu sportlichen Fahrleistungen zu verhelfen. Doch während der Motor im Audi längs eingebaut war, musste er im Golf GTI quer verbaut werden. Der Reihenmotor hatte eine oben liegende Nockenwelle, die per Zahnriemen angetrieben wurde. Pro Zylinder besaß das Aggregat zwei Ventile. Die Kurbelwelle war fünffach gelagert. Das maximale Drehmoment des Motors betrug 137 Nm bei 5000/min. Zunächst war der VW Golf GTI mit einem Viergang-Getriebe ausgestattet. Nach den Werksferien 1979 verfügte er über fünf Vorwärtsgänge. Ab 1982 wuchs der Hubraum des GTI-Motors auf 1,8 Liter. Die Leistung stieg auf 82 kW (112 PS) bei 5.800/min.

The engine

Engine specifications

Type:	Four-stroke in-line engine
Number of cylinders:	4
Displacement:	1588 cm³
Bore:	79.5 mm
Stroke:	80 mm
Power:	81 kW (110 hp)
	at 6100 rpm

The heart of the GTI originated in Ingolstadt, where the engine technicians had equipped the 1.6-liter 4-cylinder engine of the Audi 80 GT with a Bosch K-Jetronic. With this gasoline injection, the engine provided 81 kW (110 hp) at 6100 rpm. These data made the Audi engine the device of choice to get sporty driving characteristics out of the VW Golf. While the engine was mounted lengthwise in the Audi, it had to be installed across in the Golf GTI. The in-line engine had an overhead camshaft driven by a toothed belt and two valves per cylinder. The crankshaft was suspended at five points. The maximum torque amounted to 137 Nm at 5000 rpm.

At first, the VW Golf GTI was equipped with a four-gear transmission. After the plant vacation shutdown in 1979, the car had five forward gears. From 1982 on, the engine displacement was increased to 1.8 liters, thus increasing the power to 82 kW (112 hp) at 5800 rpm.

Ein Golfball als Schaltknauf und das legendäre Spucknapf-Lenkrad prägten die Optik des Cockpits.

A gear shift knob in the shape of a golf ball and the legendary "spittoon" steering wheel gave the cockpit its distinctive appearance.

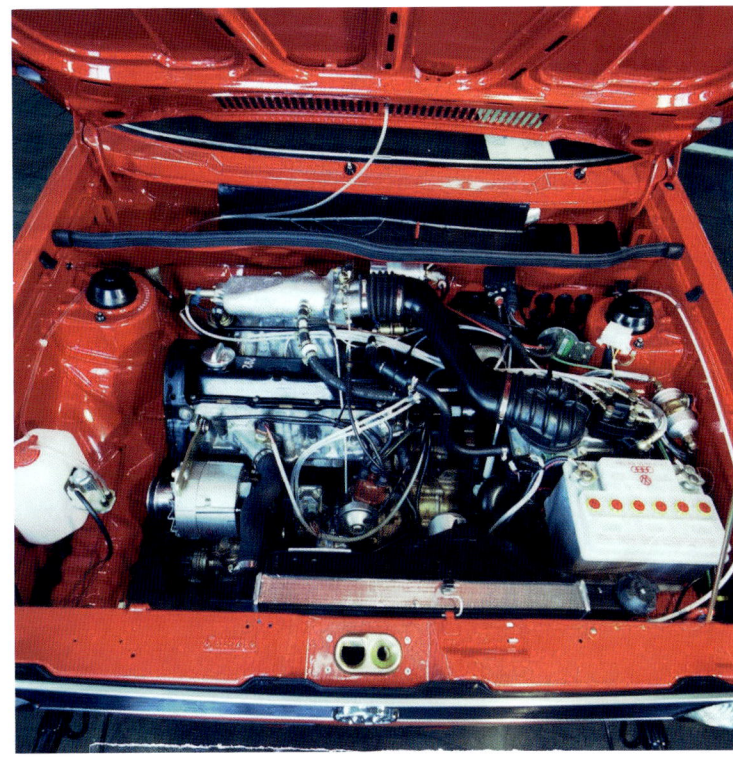

Der 1,6-Liter-Vierzylinder-Motor war in Ingolstadt für den Audi 80 GTE entwickelt worden.

The 1.6-liter 4-cylinder engine had been developed in Ingolstadt for the Audi 80 GTE.

Die sportliche Ausgabe des Käfer-Nachfolgers übertraf alle Verkaufserwartungen des Wolfsburger Managements. Zwischen 1976 und 1983 liefen bei Volkswagen insgesamt 270.000 Exemplare des Golf GTI vom Band.

The sports version of the Beetle successor exceeded all sales expectations of the VW management. Between 1976 and 1983, 270,000 units of the Golf GTI left the assembly line.

Zwischen 1978 und 1981 baute BMW insgesamt 445 Exemplare des Mittelmotor-Sportwagens M1.
From 1978 to 1981, BMW built 445 units of the midship engine sports car M1.

BMW M1

Fahrzeugdaten

Hersteller:	BMW
Land:	Deutschland
Modell:	M1
Bauzeit:	1978–1981
Länge:	4.360 mm
Breite:	1.824 mm
Höhe:	1.140 mm
Radstand:	2.560 mm
Leergewicht:	1.300 kg
Antriebsart:	Hinterrad
Höchstgeschwindigkeit:	261 km/h
Verbrauch:	19,6 Liter/100 km

Der Anstoß für den Bau des Mittelmotor-Sportwagens mit dem Projektcode „E26" kam aus dem Rennsport: BMW-Motorsportchef Jochen Neerpasch musste 1976 erleben, dass seine Autos bei den Rennen um die Marken-Weltmeisterschaft gegen die Rivalen von Porsche chancenlos waren. Eine Neukonstruktion musste her, wenn BMW gegen die Zuffenhausener Turbo-Boliden siegen wollte.

Damit aber ein solcher Rennwagen für die WM entwickelt werden konnte, stand zunächst eine Version für den normalen Straßenverkehr auf dem Programm. Das Reglement schrieb vor, dass innerhalb von 24 Monaten zunächst mindestens 400 Serien-Exemplare auf die Räder gestellt werden mussten. Im Herbst 1978 wurde die geforderte Straßenversion vorgestellt: Der BMW M1 war 277 PS stark, 264 km/h schnell und kostete 100.000 DM. Das ganze Projekt aber litt unter Problemen und Verzögerungen. Erst 1981 ging eine Gruppe-5-Rennversion des M1 in der Marken-WM an den Start – und feierte nur einen einzigen Sieg.

Specifications

Manufacturer:	BMW
Country:	Germany
Model:	M1
Produced:	1978–1981
Length:	4360 mm
Width:	1824 mm
Height:	1140 mm
Wheelbase:	2560 mm
Empty weight:	1300 kg
Type of drive:	Rear-wheel drive
Max. speed:	261 km/h
Fuel consumption:	19.6 l/100 km

The construction of this midship engine sports car with the project code "E26" was suggested by events in motorsports: in 1976, BMW motorsports director Jochen Neerpasch had to witness they did not stand a chance against the Porsche rivals at the brand world championship. A new design was needed when BMW wanted to fight the turbo racers from Zuffenhausen.

However, they first needed a streets version so that they could develop a racing car for the world championship. According to the rules, at least 400 production model units had to be built within 24 months. In autumn 1978, the required streets version was introduced. The BMW M1 had 277 hp, a top speed of 264 km/h and cost 100,000 DM. However, the project was riddled with problems and delays. The Group 5 racing version of M1 participated in the brand championship no sooner than in 1981, and it only achieved one win.

Der Motor

Motordaten

Bauart:	Viertakt-Reihenmotor
Zylinderzahl:	6
Hubraum:	3.453 cm³
Bohrung:	93,4 mm
Hub:	84 mm
Leistung:	204 kW (277 PS)
	bei 6.500/min

Für das Serienmodell des BMW M1 wurde ein neuer Sechszylinder-Reihenmotor mit 3,5 Litern Hubraum entwickelt. Er trug die werksinterne Typenbezeichnung „M88". Der Motor basierte auf dem bewährten Sechszylinder-Serientriebwerk und besaß den Vierventil-Zylinderkopf der CSL-Rennmotoren. Zudem verfügte das Antriebsaggregat des M1 über drei Doppel-Drosselklappenstutzen mit sechs Einzeldrosselklappen, eine mechanische Kugelfischer-Einspritzanlage und ein elektronisches Zündsystem. Zusätzlich statteten die Motorentechniker den M88 mit einer Trockensumpfschmierung aus. Seine maximale Leistung von 204 kW (277 PS) gab der Reihen-Sechszylinder bei 6.500/min ab. Das reichte, um den Mittelmotor-Sportler in 5,6 Sekunden aus dem Stand auf Tempo 100 zu beschleunigen.

1981 endete die Produktion des M1 nach nur 445 Exemplaren. Doch die Antriebsquelle des bayerischen Sportwagens erlebte ab 1984 ein Comeback: BMW verbaute das Kraftpaket auch im M635 CSi Coupé und im M5.

The engine

Engine specifications

Type:	Four-stroke in-line engine
Number of cylinders:	6
Displacement:	3453 cm³
Bore:	93.4 mm
Stroke:	84 mm
Power:	204 kW (277 hp)
	at 6500 rpm

For the production model of the BMW M1, a new 6-cylinder in-line engine with a displacement of 3.5 liters was developed. Its internal type designation was "M88." This engine was based on the proven 6-cylinder production engine and had the four-valve cylinder head of the CSL racing engines. Additional, the M1 engine was equipped with three twin throttle housings with six individual throttle valves, mechanical Kugelfischer injection, electronic ignition and dry sump lubrication. The maximum power of 204 kW (277 hp) was reached at 6500 rpm. This was enough to accelerate the midship engine sports car from 0 to 100 km/h in 5.6 seconds.

In 1981, production of the M1 ended after only 445 units had been built. However, the engine of the Bavarian sports car had a comeback in 1984, when BMW mounted it in the M635 CSi Coupé and in the M5.

Der Reihen-Sechszylinder des M1 war der erste
BMW-Serienmotor mit Vierventil-Technik.

The 6-cylinder in-line engine of the M1 was the first
BMW production engine with four-valve technology.

Der BMW M1 war für die Rennstrecke konzipiert worden. Bekannt wurde der Sportwagen, als BMW einen Markenpokal im Rahmen von Formel-1-Rennen ausrichtete. Niki Lauda gewann die Procar-Serie 1979.

The BMW M1 was designed for racing tracks. It became famous when BMW organized a brand cup for identical cars within the scope of Formula 1 races. Niki Lauda won the Procar series in 1979.

Drei Coupés: Audi Sport Quattro, Audi Quattro und Audi Coupé GT5E (von links).
Three coupés: Audi Sport Quattro, Audi Quattro and Audi Coupé GT5E (left to right).

Im März 1980 wurde der erste Audi Quattro auf dem Genfer Auto-Salon vorgestellt.
In March 1980, the first Audi Quattro was introduced at the Geneva Motor Show.

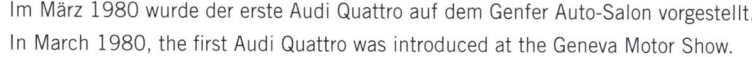

Der Innenraum des Allrad-Sportlers präsentierte sich 1980 sachlich-nüchtern.
In 1980, the sports car with all-wheel drive showed a rather sober and matter-of-fact interior.

Audi Quattro

Fahrzeugdaten

Hersteller:	Audi
Land:	Deutschland
Modell:	Quattro
Bauzeit:	1980–1991
Länge:	4.404 mm
Breite:	1.723 mm
Höhe:	1.344 mm
Radstand:	2.524 mm
Leergewicht:	1.300 kg
Antriebsart:	Allrad
Höchstgeschwindigkeit:	222 km/h
Verbrauch:	11,3 Liter/100 km

*Werte für 1980 Audi Quattro

Die VW-Tochter Audi produzierte in den 1970er-Jahren Autos, die beim Betrachter kaum für Aufregung oder Begeisterung sorgten. Das Image der Marke war bieder, die Produkte waren eher bodenständig. Doch das sollte sich im Frühjahr 1980 ändern, als der Hersteller aus dem bayerischen Ingolstadt auf dem Genfer Auto-Salon eine revolutionäre Konstruktion präsentierte: den Audi Quattro mit permanentem Allradantrieb. Das Konzept war nicht neu, denn schon in den 1960er-Jahren hatte es erste Versuche mit vierradgetriebenen Straßenfahrzeugen gegeben. Über Kleinserien waren diese Projekte jedoch nie hinausgekommen. Der Quattro mit seinem Turbomotor aber veränderte mit einem Schlag den Rallye-Sport – und bescherte Audi Siege und Weltmeistertitel. Die erste Serienversion leistete 147 kW (200 PS) und wurde später als „Ur-Quattro" bezeichnet. Ab 1989 kam ein 220 PS starker Vierventiler zum Einsatz. Bis 1991 entstanden knapp 11.500 Exemplare vom Typ Audi Quattro.

Specifications

Manufacturer:	Audi
Country:	Germany
Model:	Quattro
Produced:	1980–1991
Length:	4404 mm
Width:	1723 mm
Height:	1344 mm
Wheelbase:	2524 mm
Empty weight:	1300 kg
Type of drive:	All-wheel drive
Max. speed:	222 km/h
Fuel consumption:	11.3 l/100 km

* data for the 1980 Audi Quattro

The cars that the VW subsidiary Audi produced in the 1970s did not exactly cause excitement or rapture. The brand image was quite unsophisticated, its products were rather pragmatic. This changed in spring 1980, when the manufacturer from the Bavarian city of Ingolstadt presented a revolutionary design on the Geneva Motor Show, namely the Audi Quattro with permanent all-wheel drive. This was not a new concept, as there already had been experiments with all-wheel drives for streets cars in the 1960s. However, these projects never made it beyond small series runs. The Quattro and its turbo engine changed rally sports at a single stroke – and provided Audi with wins and world championship titles. The first series version had 147 kW (200 hp) and was later called the "original Quattro." From 1989, a four-valve engine with 220 hp was used. Until 1991, nearly 11,500 Audi Quattro were built.

Der Motor

Motordaten

Bauart:	Viertakt-Reihenmotor mit Turbolader
Zylinderzahl:	5
Hubraum:	2.144 cm³
Bohrung:	79,5 mm
Hub:	86,4 mm
Leistung:	147 kW (200 PS) bei 5.500/min

Der Audi Quattro wurde von einem Fünfzylinder-Turbo-Reihenmotor mit 2,1 Litern Hubraum angetrieben. Das wassergekühlte Triebwerk war mit einem KKK-Abgas-Turbolader und Ladeluftkühlung ausgestattet. Zudem verfügte der Audi-Motor über eine mechanische Benzineinspritzung, zwei Ventile pro Zylinder und eine oben liegende Nockenwelle. Sein Verdichtungsverhältnis betrug 7:1. Das Antriebsaggregat des Quattro leistete 147 kW (200 PS) bei 5.500/min. Bei 3.500 Touren erreichte der Fünfzylinder sein maximales Drehmoment von 285 Nm. Er beschleunigte den Ur-Quattro aus dem Stand innerhalb von 7,1 Sekunden auf Tempo 100. Die Höchstgeschwindigkeit des 1980 eingeführten Topmodells der Ingolstädter betrug 222 km/h.

Für den Mythos des Audi Quattro sorgte vor allem der permanente Allradantrieb. Dieses Konzept bescherte dem Coupé aus dem Volkswagen-Konzern eine überlegene Traktion bei allen Straßenverhältnissen und sorgte für zahlreiche Erfolge auf den Rallye-Pisten.

The engine

Engine specifications

Type:	Four-stroke in-line engine with turbo charger
Number of cylinders:	5
Displacement:	2144 cm³
Bore:	79.5 mm
Stroke:	86.4 mm
Power:	147 kW (200 hp) at 5500 rpm

The Audi Quattro was driven by a 5-cylinder in-line turbo engine with a displacement of 2.1 liters. The water-cooled engine was equipped with a KKK exhaust turbo charger and charge air cooling. Furthermore, the Audi engine had mechanical fuel injection, two valves per cylinder and an overhead camshaft. Its compression ratio amounted to 7:1. It delivered 147 kW (200 hp) at 5500 rpm and reached its maximum torque of 285 Nm at 3500 rpm. Acceleration of the original Quattro from 0 to 100 km/h took 7.1 seconds. The top speed of the top model introduced in 1980 amounted to 222 km/h.

It was particularly the permanent all-wheel drive that contributed to the legend of the Audi Quattro. This design gave the coupé of the Volkswagen group a superior traction in all road conditions and allowed many successes on rally tracks.

Durchleuchtet: Ein Blick auf die Antriebstechnik des Ingolstädter Coupés.

X-ray: A look at the drive technology of the coupé from Ingolstadt.

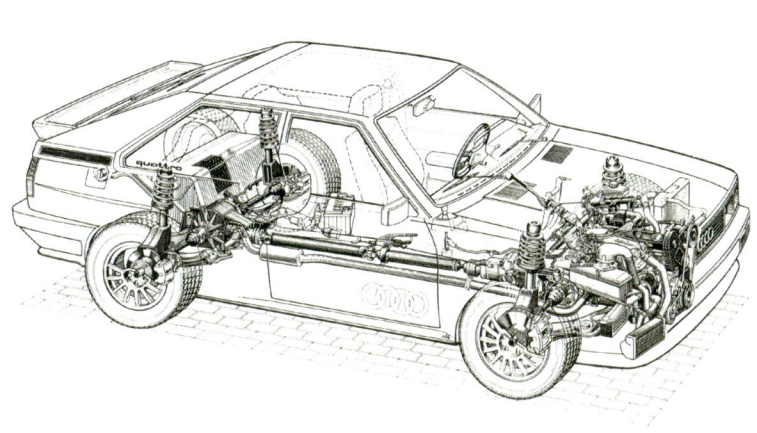

Audi setzte 1980 im Quattro auf einen Reihen-Fünfzylinder-Motor mit Turbolader.

For the Quattro, Audi relied on a 5-cylinder in-line engine with turbo charger.

Der Quattro veränderte ab 1980 das Image der Marke Audi und revolutionierte den Rallye-Sport. Die Ingolstädter Konstruktion war der Konkurrenz klar überlegen. Wer Weltmeisterschafts-Siege wollte, musste in der Folge auch auf Allradantrieb setzen.

The 1980 Quattro changed the image of the Audi brand and revolutionized the rally sports. The design from Ingolstadt was clearly superior. Whoever wanted to win in a world championship also had to use an all-wheel drive.

Nissan 300ZX Turbo von 1988.
Nissan 300ZX Turbo of 1988.

Nissan 300ZX

Fahrzeugdaten

Hersteller:	Nissan
Land:	Japan
Modell:	300ZX
Baujahr:	1983–2000
Länge:	4.405 mm
Breite:	1.725 mm
Höhe:	1.295 mm
Radstand:	2.320 mm
Leergewicht:	1.440 kg
Antriebsart:	Hinterrad
Höchstgeschwindigkeit:	224 km/h
Verbrauch:	14,0 Liter/100 km

Die Z-Reihe von Nissan zählt zu den erfolgreichsten Sportwagenserien der Welt und kann bis heute auf rund 1,7 Millionen verkaufte Einheiten zurückblicken. Ab 1983 wurde mit dem 300ZX die dritte Generation produziert; sie wurde anfangs unter dem Markennamen Datsun vertrieben, ab Ende 1985 mit dem Nissan-Logo. Der 300ZX war der erste Nissan-Sportwagen mit einem V6-Motor. Seine kompakte Bauweise erlaubte es den Designern, die Frontpartie des keilförmigen Flitzers weit nach unten zu ziehen. Sie wurde wegen ihrer halb abgedeckten Scheinwerfer zum unverkennbaren Kennzeichen des Japaners. Dank seiner aerodynamischen Formgebung erreichte die Turbo-Version des 300ZX einen beachtenswerten cw-Wert von 0,30.

In seiner Urform wurde der 300ZX bis 1989 angeboten, in einer überarbeiteten Version überzeugte er mit stärkeren Motoren und reichhaltiger Ausstattung. Dazu zählte etwa die von Nissan entwickelte HICAS-Allradlenkung.

Specifications

Manufacturer:	Nissan
Country:	Japan
Model:	300ZX
Produced:	1983–2000
Length:	4405 mm
Width:	1725 mm
Height:	1295 mm
Wheelbase:	2320 mm
Empty weight:	1440 kg
Type of drive:	Rear-wheel drive
Max. speed:	224 km/h
Fuel consumption:	14.0 liters/100 km

The Z series of Nissan is one of the world's most successful sports car series. Up to now, more than 1.7 million units were sold. Beginning with 1983, the 300ZX was produced as the third generation of this series. First, it was sold under the Datsun brand; in the end of 1985, the Nissan logo was used. The 300ZX was the first Nissan sports car with a V6 engine. Its compact construction allowed the designers to pull the front part of the wedge-shaped racer further down. This front part with the half-covered headlamps became the distinguishing mark of the Japanese car. Thanks to its aerodynamic shape, the turbo version of the 300ZX had a respectable air drag coefficient of 0.30.

The original version of the 300ZX was sold until 1989. A new modified version could score with more powerful engines and richer equipment, e.g. the HICAS all-wheel steering developed by Nissan.

Der Motor

Motordaten

Bauart:	Viertakt-V6-Ottomotor
Zylinderzahl:	6
Hubraum:	2.960 cm³
Bohrung:	87 mm
Hub:	83 mm
Leistung:	149 kW (203 PS)
	bei 5.200/min

Im Nissan 300ZX fand der erste japanische V6-Motor seinen Einsatz. Der Motor mit der Typenbezeichnung „VG30" war ein Sechszylinder-V-Motor in 60-Grad-Bauweise. Er holte aus seinen 3 Litern Hubraum als Saugversion 119 kW (162 PS). Mit dem ab 1985 erhältlichen Turbolader brachte es der Sportwagen sogar auf 149 kW (203 PS) bei 5.200 Umdrehungen pro Minute. Mit ihm erreichte die Nissan-Z-Serie erstmals die 250-km/h-Marke. Der wassergekühlte Motor erwies sich als wahres Kraftpaket, was durch sein beachtliches Drehmoment von 309 Nm bei 3.600 Umdrehungen pro Minute unterstrichen wurde. Das Verdichtungsverhältnis der V6-Maschine lag bei 7,8:1. Das Aggregat besaß zwei Ventile pro Zylinder, die über oben liegende Nockenwellen angetrieben wurden. Der Motor wartete ferner mit Hydrostößel mit automatischem Ventilspielausgleich, Schubabschaltung und Transistorzündung auf. Mit 162 Kilogramm zählte das VG30-Aggregat zudem zu den Motor-Leichtgewichten.

The engine

Engine specifications

Type:	4-stroke V6 Otto engine
Number of cylinders:	6
Displacement:	2960 cm³
Bore:	87 mm
Stroke:	83 mm
Power:	149 kW (203 hp)
	at 5200 rpm

The Nissan 300ZX used the first Japanese V6 engine. It was a 6-cylinder V engine with a bank angle of 60°, labeled as "VG30." The suction version produced 119 kW (162 hp) from a displacement of 3 liters. With the turbo charger first available in 1985, the sports car even achieved 149 kW (203 hp) at 5200 rpm. With this engine, the Nissan Z series reached the 250 km/h mark for the first time. The water-cooled engine proved to be a real power pack. This was also made evident by the respectable torque of 309 Nm at 3600 rpm. The compression ratio amounted to 7.8:1. The engine had two valves per cylinder, which were driven by overhead camshafts. Furthermore, the engine boasted hydraulic valve lifters with automatic valve-clearance compensation, throttle cutoff and transistor ignition. At 162 kg, the VG30 is also a lightweight among engines.

Motor des Nissan 300ZX Turbo von 1988.

Engine of the Nissan 300ZX Turbo of 1988.

V6-Motor des Nissan 300ZX.

V6 engine of the Nissan 300ZX.

Der BMW 750i von 1987 war der erste deutsche Nachkriegswagen mit einem 12-Zylinder-Motor.
The BMW 750i of 1987 was the first German post-war car with a 12-cylinder engine.

BMW 750i

Fahrzeugdaten

Hersteller:	BMW
Land:	Deutschland
Modell:	750i
Bauzeit:	1987–1994
Länge:	4.910 mm
Breite:	1.845 mm
Höhe:	1.400 mm
Radstand:	2.833 mm
Leergewicht:	1.800 kg
Antriebsart:	Hinterrad
Höchstgeschwindigkeit:	250 km/h
Verbrauch:	13,3 Liter/100 km

Der 1987 auf dem Genfer Auto-Salon vorgestellte BMW 750i war ein Meilenstein der deutschen Automobil-Historie: Als erstes deutsches Serienautomobil nach dem Zweiten Weltkrieg besaß die Neukonstruktion aus München einen Zwölfzylinder-Motor. Damit platzierte BMW die Luxus-Limousine in einem Marktsegment, das zuvor ausschließlich Edelmarken wie Ferrari oder Jaguar vorbehalten war. Die Kundschaft zeigte sich begeistert: Noch bevor der 750i seine offizielle Premiere feierte, konnten die Verkäufer mehr als 3.000 Vorbestellungen notieren.

Der BMW 750i überzeugte mit einer Mischung aus Exklusivität, Komfort und Sportlichkeit. Der V12-Zylinder-Motor leistete mit fünf Litern Hubraum 220 kW (300 PS). Für den Sprint von 0 auf Tempo 100 benötigte das weiß-blaue Flaggschiff 7,5 Sekunden. Die Höchstgeschwindigkeit limitierten die BMW-Techniker auf 250 km/h – auch mit Rücksicht auf den damaligen Stand der Reifentechnik. Ein Top Speed von 270 km/h wäre möglich gewesen.

Specifications

Manufacturer:	BMW
Country:	Germany
Model:	750i
Produced:	1987–1994
Length:	4910 mm
Width:	1845 mm
Height:	1400 mm
Wheelbase:	2833 mm
Empty weight:	1800 kg
Type of drive:	Rear-wheel drive
Max. speed:	250 km/h
Fuel consumption:	13.3 l/100 km

The BMW 750i introduced at the Geneva Motor Show in 1987 was a milestone of the German automobile history. The new design from Munich was the first German production car after World War II with a 12-cylinder engine. In this way, BMW positioned the limousine in a market segment that had been the exclusive domain of noble brands like Ferrari and Jaguar. The customers were enthusiastic. Even before the official debut of the 750i, dealers could register more than 3000 pre-orders.

The BMW 750i boasted with a mixture of exclusiveness, comfort and sportiveness. The V12 engine with a displacement of 5 liters provided 220 kW (300 hp). Accelerating from 0 to 100 km/h took the flagship of the Bavarian manufacturer no more than 7.5 seconds. With regard to the state of the contemporary tire technology, the BMW engineers limited the top speed to 250 km/h, although 270 km/h would have been possible.

Der Motor

Motordaten

Bauart:	Viertakt-60-Grad-V-Motor
Zylinderzahl:	12
Hubraum:	4.988 cm³
Bohrung:	84 mm
Hub:	75 mm
Leistung:	220 kW (300 PS)
	bei 5.200/min

Schon lange vor der Premiere des 750i im Jahr 1987 hatten sich die BMW-Motorentwickler mit dem Thema Serien-Zwölfzylinder beschäftigt: 1972 begann die Arbeit unter dem Projektnamen „M33". Doch als dieses erste 5,0-Liter Triebwerk 1974 einsatzfähig war, brachte es 315 Kilogramm auf die Waage. Damit war der V12 deutlich zu schwer. Eine Neuentwicklung mit dem Namen „M66" fiel 1977 zwar um 40 Kilogramm leichter aus, doch auch dieses Projekt wurde eingestellt. Im November 1982 schließlich nahmen die BMW-Entwickler einen dritten Anlauf für einen V12-Serienmotor. Sie entschieden sich für eine komplette Neukonstruktion. Und im Oktober 1983 absolvierte der M70 seinen ersten Prüfstandlauf. Der V12-Motor verfügte über 5,0 Liter Hubraum und leistete 220 kW (300 PS) bei 5.200/min. Um Gewicht zu sparen, setzten die Konstrukteure auf ein Kurbelgehäuse aus Aluminium. Der M70 brachte es auf ein Komplettgewicht von nur 240 Kilogramm. Ab 1989 trieb der Motor auch den 850i an.

The engine

Engine specifications

Type:	Four-stroke 60° V engine
Number of cylinders:	12
Displacement:	4988 cm³
Bore:	84 mm
Stroke:	75 mm
Power:	220 kW (300 hp)
	at 5200 rpm

Long before the debut of the 750i in 1987, the BMW engine designers had explored the topic of 12-cylinder engines for production cars. The work began in 1972, code-named "M33." However, the first 5.0-liter V12 engine that was ready for use in 1974 weighed 315 kg and was thus much too heavy. In 1977, a new development called "M66" weighed 40 kg less, but eventually, this project was also cancelled. In November 1982, the BMW developers started a new attempt for a V12 production engine. This time they opted for a complete re-design. In October 1983, the M70 completed its first test run. This V12 engine had a displacement of 5.0 liters and provided 220 kW (300 hp) at 5200 rpm. In order to save weight, the designers relied on an aluminum crankcase. The total weight of the M70 amounted to only 240 kg. Beginning with 1989, this engine also powered the 850i.

Ab 1972 gab es bei BMW immer wieder
Pläne für einen Zwölfzylinder-Motor.
Beginning with 1972, BMW came up with
various plans for a 12-cylinder engine.

Das Triebwerk mit der internen Bezeichnung
„M70" wurde auch im BMW 850i eingesetzt.
The engine with the internal designation
"M70" was also used in the BMW 850i.

1987 brachte BMW die Luxus-Limousine 750i mit einem 5,0-Liter-V12-Motor auf den Markt. Dieser erste deutsche Serien-Zwölfzylinder seit dem Zweiten Weltkrieg leistete 220 kW (300 PS) bei 5.200/min.

In 1987, BMW launched the limousine 750i with a 5.0 liter V12 engine. This was the first German series 12-cylinder engine since World War II. It delivered a power of 220 kW (300 hp) at 5200 rpm.

Beim Facelift-Modell von 2002 verschwanden die typischen Klappscheinwerfer.
The typical pop-up headlights disappeared in the model facelift of 2002.

Honda NSX

1990

Fahrzeugdaten

Hersteller:	Honda
Land:	Japan
Modell:	NSX
Bauzeit:	1990–2005
Länge:	4.405 mm
Breite:	1.810 mm
Höhe:	1.170 mm
Radstand:	2.530 mm
Leergewicht:	1.370 kg
Antriebsart:	Hinterrad
Höchstgeschwindigkeit:	270 km/h
Verbrauch:	ca. 12,5 Liter/100 km

*Werte für 1990 Honda NSX

Der Honda NSX war ein Mittelmotor-Sportwagen, den der japanische Hersteller ab 1990 produzierte. Er war mit einem 3,0-Liter-V6-Triebwerk ausgestattet, das 201 kW (274 PS) leistete. Damit sollte der NSX speziell auf dem wichtigen US-Markt den traditionsreichen europäischen Sportwagen-Herstellern wie Porsche und Ferrari Konkurrenz machen. Als entsprechendes Signal wurde bereits der Ort der Premiere gewertet: Honda stellte den NSX erstmals 1989 auf der Chicago Auto Show in den USA vor. Rund die Hälfte der knapp 9.000 gebauten Exemplare wurden später in den USA unter dem Namen „Acura NSX" verkauft.

In den 1980er Jahren war Honda als Motorenlieferant sehr erfolgreich in der Formel 1 aktiv. So gewann Ayrton Senna alle seine drei WM-Titel in Fahrzeugen mit Honda-Motoren. Bei der Entwicklung des NSX unterstützte Senna die Honda-Ingenieure. Der Brasilianer absolvierte 1990 Abstimmungsfahrten mit dem Sportwagen auf der Nordschleife des Nürburgrings.

Specifications

Manufacturer:	Honda
Country:	Japan
Model:	NSX
Produced:	1990–2005
Length:	4405 mm
Width:	1810 mm
Height:	1170 mm
Wheelbase:	2530 mm
Empty weight:	1370 kg
Type of drive:	Rear-wheel drive
Max. speed:	270 km/h
Fuel consumption:	approx. 12.5 l/100 km

* data for the 1990 Honda NSX

The Honda NSX was a sports car with midship engine built by the Japanese manufacturer from 1990 on. It was equipped with a 3.0-liter V6 engine with 201 kW (274 hp) to compete with traditional European sports car manufacturers like Porsche and Ferrari, in particular on the important US market. The location of the debut was deemed as an apt signal, as Honda introduced the NSX at the Chicago Auto Show in 1989. Approx. half of the nearly 9000 units were sold in the USA as Acura NSX. In the 1980s, Honda was a very successful engine supplier for Formula 1. For instance, Ayrton Senna won all of his three world championship titles in cars with Honda engines. He also supported the Honda engineers during the development of the NSX. In 1990, the Brazilian driver carried out the calibration drives for this sports car on the North Loop of the Nürburgring.

Der Motor

Motordaten

Bauart:	Viertakt-90-Grad-V-Motor
Zylinderzahl:	6
Hubraum:	2.977 cm³
Bohrung:	90 mm
Hub:	78 mm
Leistung:	201 kW (274 PS)
	bei 7.100/min

Honda stattete den NSX mit einem V6-Saugmotor aus, der 201 kW (274 PS) bei 7.100/min leistete. Seine zulässige Höchstdrehzahl lag bei 8.000 Umdrehungen pro Minute. Bei 5.400/min erreichte der Sechszylinder sein maximales Drehmoment von 285 Nm. Jeder Zylinder war mit zwei Ein- und zwei Auslassventilen versehen. Der Motor besaß zudem eine variable Ventilsteuerung. Das Verdichtungsverhältnis der wassergekühlten Maschine lag bei 10,2:1. Serienmäßig besaß der NSX ein Fünfgang-Schaltgetriebe. In dieser Ausführung beschleunigte der Sportwagen in 5,9 Sekunden von 0 auf 100 km/h und erreichte eine Spitzengeschwindigkeit von 270 km/h. Auf Wunsch war der Wagen auch mit einer Viergang-Automatik und 188 kW (256 PS) erhältlich. In den Autos der zweiten Generation des NSX kam ab 1997 ein 3,2-Liter-V6-Motor mit 205 kW (280 PS) zum Einsatz. Honda stattete den Wagen zudem mit einem Sechsgang-Schaltgetriebe aus. Die Produktion dieses Modells endete 2005.

The engine

Engine specifications

Type:	Four-stroke 90° V engine
Number of cylinders:	6
Displacement:	2977 cm³
Bore:	90 mm
Stroke:	78 mm
Power:	201 kW (274 hp)
	at 7100 rpm

Honda equipped the NSX with a V6 suction engine with 201 kW (274 hp) at 7100 rpm. Its maximum engine speed amounted to 8000 rpm. At 5400 rpm, it reached its maximum torque of 285 Nm. Each cylinder had two intake and two outlet valves. Furthermore, the engine was equipped with variable valve control. The compression ratio of the water-cooled engine amounted to 10.2:1. The NSX had a manual five-gear transmission in base. In this version, the sports car accelerated from 0 to 100 km/h in 5.9 seconds and reached a top speed of 270 km/h. Optionally, a four-gear automatic transmission and an engine with 188 kW (256 hp) were available. Beginning with 1997, the cars of the second generation had a 3.2-liter V6 engine with 205 kW (280 hp). Honda also put a six-gear transmission in the car. Production of this model ended in 2005.

Der Honda NSX NA1 aus dem Jahr 1990
wurde von einem 3,0-Liter-V6-Motor
angetrieben.
The 1990 Honda NSX NA1 was driven
by a 3.0-liter V6 engine.

Mit dem Mittelmotor-Modell NSX wollte Honda den traditionsreichen Sport-wagen-Herstellern wie Porsche ab 1990 Konkurrenz machen. Die meisten Fahrzeuge setzten die Japaner auf dem US-Markt ab.

Beginning in 1990, Honda used the midship engine model NSX to compete with traditional sports car manufacturers like Porsche. Most of the cars were sold in the USA.

Der Lamborghini Diablo VT wurde ab 1993 angeboten. Sein besonderes Merkmal war der Allradantrieb.
The Lamborghini Diablo VT was available beginning with 1993. As a special feature, it came with an all-wheel drive.

Lamborghini Diablo VT

Fahrzeugdaten

Hersteller:	Lamborghini
Land:	Italien
Modell:	Diablo VT
Baujahr:	1993–1999
Länge:	4.458 mm
Breite:	2.040 mm
Höhe:	1.105 mm
Radstand:	2.670 mm
Leergewicht:	1.625 kg
Antriebsart:	Allrad
Höchstgeschwindigkeit:	326 km/h
Verbrauch:	ca. 27 Liter/100 km

Mit dem von 1990 bis 2001 gebauten Diablo verfolgte Lamborghini das Ziel, das schnellste Serienauto der Welt zu bauen. Seine Höchstgeschwindigkeit wurde mit 325 km/h angegeben. Dieser Wert wurde bei unabhängigen Tests mehrfach überboten. In seiner Grundausstattung wurde der Diablo nur mit Hinterradantrieb ausgeliefert. Ab 1993 wurde die Modellreihe um den „VT" (Visco Traction) mit Allradantrieb erweitert. Dieser bot noch exzellentere Fahreigenschaften, ohne dass man Einbußen an der erreichbaren Höchstgeschwindigkeit hinnehmen musste. Für die Form des Diablo zeichnete der Fahrzeugdesigner Marcello Gandini verantwortlich. Er verlieh dem Wagen schlichte Eleganz. Zu den Designmerkmalen zählten neben den großzügigen seitlichen Lufteinlässen die bis heute für Lamborghini typischen Scherentüren.

Bei der Karosserie kam eine Kombination aus Leichtmetall, Stahl, Karbonfaserverstärktem Kunststoff und weiteren exklusiven Materialien zum Einsatz.

Specifications

Manufacturer:	Lamborghini
Country:	Italy
Model:	Diablo VT
Produced:	1993–1999
Length:	4458 mm
Width:	2040 mm
Height:	1105 mm
Wheelbase:	2670 mm
Empty weight:	1625 kg
Type of drive:	All-wheel drive
Max. speed:	326 km/h
Fuel consumption:	approx. 27 liters/100 km

With the Diablo built from 1990 to 2001, Lamborghini aimed at building the fastest series car in the world. Its maximum speed was rated at 325 km/h, but this figure was even surpassed several times in independent tests. In the standard version, the Diablo was only shipped with rear-wheel drive. Beginning with 1993, the series was extended by the VT (Visco Traction) all-wheel drive. It offered even better driving characteristics without sacrificing the achievable maximum speed. The credit for the appearance of the Diablo goes to the vehicle designer Marcello Gandini who endowed the car with simple elegance. In addition to the ample air intakes at the side, the design characteristics also comprise the scissor doors that are still signature features of Lamborghini.

The autobody was made of a combination of light metal, steel, carbon fiber reinforced plastic and other exclusive materials.

Der Motor

Motordaten

Bauart:	Otto-V-Motor
Zylinderzahl:	12
Hubraum:	5.707 cm³
Bohrung:	87 mm
Hub:	80 mm
Leistung:	367 kW (499 PS)
	bei 7.000/min

In allen Lamborghini Diablos sorgte ein mächtiger 12-Zylinder-V-Motor mit einem Gabelwinkel von 60 Grad für Kraft, Power und Dynamik. Das wassergekühlte Aggregat des Diablo VT von 1993 hatte einen Hubraum von 5,7 Litern und brachte es auf beachtliche 367 kW (499 PS), die bei 7.000 Umdrehungen pro Minute erreicht wurden. Sein höchstes Drehmoment von 580 Nm stellte die Maschine bei 5.200 Umdrehungen pro Minute zur Verfügung. Das Verdichtungsverhältnis des Motors lag bei 10:1. Seine vier Ventile pro Zylinder wurden über zwei oben liegende Nockenwellen je Zylinderreihe angesteuert.

Im Laufe der folgenden Jahre wurde die Leistung des Motors weiter gesteigert. In seiner Version für den 1999er Diablo VT brachte er es bei gleichem Hubraum auf 390 kW (530 PS). Zusätzlich war der VT ab 1999 mit einem V12-6-Liter-Motor mit 404 kW (550 PS) bei 7.100 Umdrehungen pro Minute verfügbar.

The engine

Engine specifications

Type:	Otto V engine
Number of cylinders:	12
Displacement:	5707 cm³
Bore:	87 mm
Stroke:	80 mm
Power:	367 kW (499 hp)
	at 7000/min

Power and dynamic of all Lamborghini Diablos were provided by a powerful 12-cylinder V engine with a bank angle of 60°. The water-cooled engine of the Diablo VT of 1993 had a displacement of 5.7 liters and achieved remarkable 367 kW (499 hp) at 7000 rpm. The maximum torque of 580 Nm was achieved at 5200 rpm. The compression ratio amounted to 10:1. The four valves per cylinder were controlled by two overhead camshafts per cylinder bank.

In the following years, the engine power was increased. The version for the 1999 model of the Diablo VT had the same displacement but provided 390 kW (530 hp). Beginning with 1999, the VT was also available with a 6-liter V12 engine with 404 kW (550 hp) at 7100 rpm.

Der 5,7-Liter-Motor des 1993er Diablo VT brachte es auf 367 kW.

The 5.7 liter engine of the 1993 Diablo VT provided 367 kW.

Der Alfa Romeo Spider 916 kam 1994 auf den Markt und verfügte über einen Frontantrieb.

The Alfa Romeo Spider 916 with front-wheel drive was launched in 1994.

Alfa Romeo Spider 916

Fahrzeugdaten

Hersteller:	Alfa Romeo
Land:	Italien
Modell:	Spider (916)
Bauzeit:	1994–2005
Länge:	4.285 mm
Breite:	1.780 mm
Höhe:	1.315 mm
Radstand:	2.540 mm
Leergewicht:	1.370 kg
Antriebsart:	Vorderrad
Höchstgeschwindigkeit:	210 km/h
Verbrauch:	9,2 Liter/100 km

*Werte für 1994 Alfa Romeo Spider 2.0 TS

Die Geschichte des Alfa Romeo Spider begann im Frühjahr 1966, als das zweisitzige Cabrio auf dem Genfer Auto-Salon sein Debüt feierte. Das Design aus dem Hause Pininfarina erwies sich als ausgesprochen langlebig: Zwar wurde der Alfa Romeo im Laufe der Jahre immer wieder leicht überarbeitet, doch am Grundkonzept änderte sich bis 1993 nichts. Erst dann stellten die Italiener die Produktion ein. Doch ein Jahr später präsentierten sie einen völlig neuen Spider.

Dieser neue Wagen mit der internen Bezeichnung „916" hatte weder formal noch technisch Ähnlichkeit mit seinem Vorgänger. Pininfarina hatte einen Roadster mit ausgeprägter Keilform auf die Räder gestellt. Zeitgleich mit dem Spider brachte Alfa Romeo eine geschlossene Coupé-Variante unter der Bezeichnung „GTV" auf den Markt. Beide verfügten über vorne quer liegende Motoren mit Frontantrieb. Die Kunden konnten zwischen Vier- und Sechszylinder-Triebwerken wählen. 2005 nahm Alfa Romeo den 916 aus dem Programm.

Specifications

Manufacturer:	Alfa Romeo
Country:	Italy
Model:	Spider (916)
Produced:	1994–2005
Length:	4285 mm
Width:	1780 mm
Height:	1315 mm
Wheelbase:	2540 mm
Empty weight:	1370 kg
Type of drive:	Front-wheel drive
Max. speed:	210 km/h
Fuel consumption:	9.2 l/100 km

* data for the 1994 Alfa Romeo Spider 2.0 TS

The history of the Alfa Romeo Spider began in spring 1966, when the two-seater cabriolet debuted at the Geneva Motor Show. The design by Pininfarina proved to be very extraordinarily long-lived. The Alfa Romeo was slightly modified in the course of the years, but the basic design did not change until 1993, when production ceased. However, in the following year, the Italian manufacturers presented a completely new Spider.

This new car with the internal designation 916 bore no resemblance to its predecessor, neither in shape nor in technology. Pininfarina had created a distinctively wedge-shaped roadster. Alfa Romeo launched a closed coupé version called GTV at the same time as the Spider. Both had front engines mounted across and front-wheel drive. Customers could choose between 4- and 6-cylinder engines. In 2005, Alfa Romeo discontinued the 916.

Der Motor

Motordaten

Bauart:	Viertakt-Reihenmotor
Zylinderzahl:	4
Hubraum:	1.970 cm³
Bohrung:	83 mm
Hub:	91 mm
Leistung:	110 kW (150 PS)
	bei 6.300/min

Für den neuentwickelten Spider 916 bot Alfa Romeo verschiedene Motoren mit einem Hubraum von 1,8 bis 3,2 Liter an. Sie wurden durchweg quer eingebaut. Bei der Markteinführung 1994 war ein 2,0-Liter-Vierzylinder die Basismotorisierung des neuen Cabrios. Der Graugussblock dieser Maschine stammte aus dem Motoren-Baukasten von Fiat. Der Zylinderkopf aus Aluminium mit zwei Zündkerzen pro Brennraum war dagegen eine Eigenentwicklung der Alfa-Romeo-Ingenieure. Der Motor mit der Bezeichnung „2.0 l Twin Spark 16V" war ein wassergekühlter Reihen-Vierzylinder mit zwei oben liegenden Nockenwellen und vier Ventilen pro Zylinder. Das 2,0-Liter-Aggregat leistete 110 kW (150 PS) bei 6.300 Umdrehungen pro Minute. Das maximale Drehmoment des Motors betrug 186 Nm bei 4.000/min. Das Verdichtungsverhältnis der italienischen Konstruktion lag bei 10:1. Die Motorleistung wurde über ein Fünfgang-Schaltgetriebe auf die Vorderräder übertragen – wie bei Alfa Romeo zu dieser Zeit generell üblich.

The engine

Engine specifications

Type:	Four-stroke in-line engine
Number of cylinders:	4
Displacement:	1970 cm³
Bore:	83 mm
Stroke:	91 mm
Power:	110 kW (150 hp)
	at 6300 rpm

For the newly developed Spider 916, Alfa Romeo provided various engines with a displacement of 1.8 to 3.2 liters. All were mounted across. At the market launch in 1994, the standard engine of the new cabriolet was a 2.0-liter 4-cylinder. The grey cast iron block was part of the Fiat engine construction kit, while the aluminum cylinder head with two spark plugs per combustion chamber was an in-house development by Alfa Romeo engineers. The engine with the designation 2.0 l Twin Spark 16V was a water-cooled 4-cylinder in-line engine with two overhead camshafts and four valves per cylinder. It provided 110 kW (150 hp) at 6300 rpm and reached its maximum torque of 186 Nm at 4000 rpm. The compression rate amounted to 10:1. The engine power was transferred to the front wheels via a manual five-gear transmission, as it was then customary at Alfa Romeo.

Der 2.0 Twin Spark 16 V war einer von sechs Motoren, die Alfa Romeo für den Spider 916 anbot.

The 2.0 Twin Spark 16 V was one of six engines that Alfa Romeo provided for the Spider 916.

Alfa GTV/Spider - Motore 2.0 T. Spark 16V
2.0 T. Spark 16V engine
Moteur 2.0 T. Spark 16V
Motor 2.0 T. Spark 16V
Motor 2.0 T. Spark 16V
Motor 2.0 T. Spark 16V

Alfa Romeo

Der Alfa Romeo Spider 916 trat in den 1990er-Jahren ein schweres Erbe an. Er sollte den legendären Spider ersetzen, der ab 1966 vom Band gelaufen war. Optisch und technisch wagten die Italiener einen Neuanfang.

In the 1990s, the Alfa Romeo Spider 916 took on a difficult legacy, as it was supposed to replace the legendary Spider that had been built since 1966. Visually and technologically, the Italian manufacturer took a chance on a fresh start.

Mit dem Vantage V600 brachte Aston Marin 1998 ein 322 km/h schnelles Sportcoupé auf den Markt.

With the 1998 Vantage V600, Aston Martin launched a sports coupé with 322 km/h.

Aston Martin Vantage V600

Fahrzeugdaten

Hersteller:	Aston Martin
Land:	Großbritannien
Modell:	Vantage V600
Bauzeit:	1998–2000
Länge:	4.745 mm
Breite:	1.944 mm
Höhe:	1.330 mm
Radstand:	2.611 mm
Leergewicht:	1.970 kg
Antriebsart:	Hinterrad
Höchstgeschwindigkeit:	322 km/h
Verbrauch:	ca. 14–21 Liter/100 km

Der Aston Martin Vantage folgte 1996 auf den bis dahin gebauten Virage. Den Vantage gab es neben der Basisversion auch in drei verschiedenen Leistungsstufen unter den Bezeichnungen „V600", „Le Mans" und „Le Mans Special Edition". Während die Basisvariante über einen 5,3 Liter großen V8-Motor mit zwei Eaton-Kompressoren und 410 kW (557 PS) Leistung verfügte, brachte es die 1998 präsentierte Version „V600" dank geänderter Kompressoren, gesteigertem Ladedruck und einer modifizierten Abgasanlage auf 600 bhp, was 447 kW (608 PS) Leistung entspricht. Die letzten 40 Exemplare des Vantage hörten indes auf die Zusatzbezeichnung „Le Mans", was an den Sieg eines Aston Martin bei dem berühmten Rennen an der Sarthe im Jahr 1959 erinnern sollte. Den Aston Martin Vantage „Le Mans" gab es mit der 410 kW (557 PS) starken Basismotorisierung sowie als „Le Mans Special Edition" mit dem Triebwerk des „V600". Beiden Modellen gemeinsam waren aerodynamische Modifikationen zur Steigerung der Höchstgeschwindigkeit.

Specifications

Manufacturer:	Aston Martin
Country:	Great Britain
Model:	Vantage V600
Produced:	1998–2000
Length:	4745 mm
Width:	1944 mm
Height:	1330 mm
Wheelbase:	2611 mm
Empty weight:	1970 kg
Type of drive:	Rear-wheel drive
Max. speed:	322 km/h
Fuel consumption:	approx. 14–21 l/100 km

In 1996, the Aston Martin Vantage became the successor of the Virage. Apart from the standard version, the Vantage was also available with three different power levels called "V600," "Le Mans" and "Le Mans Special Edition." While the standard version had a 5.3-liter V8 engine with two Eaton compressors and 410 kW (557 hp), the "V600" version introduced in 1998 could boast a power of 600 bhp, which equals 447 kW (608 hp). This increase was due to modified compression, increased charging pressure and a modified exhaust system. The final 40 units of the Vantage were called "Le Mans," which alluded to the win of an Aston Martin on the Sarthe in 1959. The Aston Martin Vantage Le Mans was available with a 410 kW (557 hp) base engine and as the "Le Mans Special Edition" with the engine of the "V600." Both models had aerodynamic modifications to increase the top speed.

Der Motor

Motordaten

Bauart:	Viertakt-90-Grad-V-Motor mit Kompressor
Zylinderzahl:	8
Hubraum:	5.341 cm³
Bohrung:	100 mm
Hub:	85 mm
Leistung:	447,4 kW (608 PS) bei 6.200/min

Im Aston Martin Vantage V600 arbeitete ein 5,3-Liter großer V8-Motor mit zwei Ein- und zwei Auslassventilen je Zylinder, die über zwei oben liegende Nockenwellen gesteuert wurden. Bei 6.200 Umdrehungen pro Minute gab der 90-Grad-V8-Motor seine maximale Leistung von 447,4 kW (608 PS) ab. Um dem Aggregat diese Leistung abzuverlangen, wurde er mit zwei Eaton-Kompressoren ausgestattet, die über einen Zahnriemen angetrieben wurden. Für möglichst kühle und damit leistungsfördernde Ladeluft sorgten zwei Ladeluftkühler. Beeindruckend an diesem Triebwerk war das enorm breite Drehzahlband sowie der im Gegensatz zu Turbomotoren kräftige Antritt bereits bei niedrigen Drehzahlen. Das höchste Drehmoment von gewaltigen 814 Nm erreichte der wassergekühlte Motor bei 4.400 Umdrehungen pro Minute. Der Aston Martin Vantage V600 verfügte über ein manuelles Fünfgang-Getriebe, war hinterradgetrieben und beschleunigte aus dem Stillstand binnen 9,3 Sekunden auf 160 km/h.

The engine

Engine specifications

Type:	Four-stroke 90° V engine with compressor
Number of cylinders:	8
Displacement:	5341 cm³
Bore:	100 mm
Stroke:	85 mm
Power:	447.4 kW (608 hp) at 6200 rpm

The Aston Martin Vantage V600 was driven by a 5.3-liter 90° V8 engine with two intake and two outlet valves per cylinder, which were controlled by two overhead camshafts. At 6200 rpm, the engine delivered its maximum power of 447.4 kW (608 hp). To achieve this power, two Eaton compressors driven by a tooth belt were used. Additionally, two charge air coolers provided cool air, which also increases the power. An impressive feature of the engine was the extremely broad engine speed range and the powerful acceleration even at low engine speed, which is something that cannot be achieved with turbo engines. The water-cooled engine reached its maximum power of a whopping 814 Nm at 4400 rpm. The Aston Martin Vantage V600 had a five-gear manual transmission, rear-wheel drive and acceleration from 0 to 160 km/h in 9.3 seconds.

Im V8-Motor des Aston Martin Vantage V600 verrichteten zwei Eaton-Kompressoren und Ladeluftkühlung ihren Dienst.

The V8 engine of Aston Martin Vantage V600 was equipped with two Eaton compressors and charge air cooling.

Der Aston Martin Vantage V600 stand exemplarisch für die hohe Kunst des britischen Sportwagenbaus. Sein V8-Motor mit zwei Eaton-Kompressoren macht ihn auch in technischer Hinsicht einzigartig.

The Aston Martin Vantage V600 was an example of the fine art of British sports car manufacturing. Its V8 engine with two Eaton compressors made it also an extraordinary car from a technological perspective.

Der Aston Martin V12 Vanquish in der von 2001 bis 2007 gebauten Version.
The Aston Martin V12 Vanquish version built from 2001 to 2007.

Aston Martin V12 Vanquish

Fahrzeugdaten

Hersteller:	Aston Martin
Land:	England
Modell:	V12 Vanquish
Bauzeit:	2001–2007
Länge:	4.665 mm
Breite:	1.923 mm
Höhe:	1.318 mm
Radstand:	2.690 mm
Leergewicht:	1.820 kg
Antriebsart:	Hinterrad
Höchstgeschwindigkeit:	306 km/h
Verbrauch:	16,4 Liter/100 km

James Bond wusste, wie er am schnellsten Bösewichte jagen kann, ohne auf britische Gediegenheit zu verzichten – was im 2002 erschienenen Bond-Streifen „Stirb an einem anderen Tag" zu erleben ist. Der Aston Martin V12 Vanquish war von 2001 bis 2007 erhältlich und wurde überwiegend in Handarbeit gefertigt. Konzipiert war er als Zweisitzer, konnte aber zum 2+2-Sitzer erweitert werden. Das Sportcoupé kam serienmäßig mit Einparkhilfe, Navigationssystem, Volllederausstattung und vielem mehr. Selbst die Farbe konnte man sich aus einer Palette von mehreren tausend Farbtönen aussuchen. Der Aston Martin V12 Vanquish begeistert aber auch durch seine zurückhaltende, zeitlose Eleganz. Der Wagen protzt nicht, sondern hält sich dezent zurück. Umso mehr versetzt er ins Staunen, wenn man die 338 kW (460 PS) seines 5,9-Liter-V12-Motors mobilisiert, denn 5,0 Sekunden für den Sprint von 0 auf 100 km/h und 306 km/h Höchstgeschwindigkeit sind starke Ansagen.

Specifications

Manufacturer:	Aston Martin
Country:	England
Model:	V12 Vanquish
Produced:	2001–2007
Length:	4665 mm
Width:	1923 mm
Height:	1318 mm
Wheelbase:	2690 mm
Empty weight:	1820 kg
Type of drive:	Rear-wheel drive
Max. speed:	306 km/h
Fuel consumption:	16.4 l/100 km

James Bond knew how to hunt down the villains fast and yet without relinquishing British "class," as you can see in the 2002 Bond flick "Die Another Day." The Aston Martin V12 Vanquish was available from 2001 to 2007 and was mostly manufactured by hand. It was designed as a two-seater but could be extended to 2+2 seats. The sports coupé came with parking assistant, navigation system, leather interior and more in base. The colour could be selected from a gamut of several thousand shades. But the Aston Martin V12 Vanquish also stands out because of its restraint and ageless elegance. The car does not boast but refrains itself. The more it surprises when it deploys the 338 kW (460 hp) of its 5.9-liter V12 engine. Just 5 seconds for accelerating from 0 to 100 km/h and a top speed of 306 km/h are really impressive.

Der Motor

Motordaten

Bauart:	Viertakt-60-Grad-V-Motor
Zylinderzahl:	12
Hubraum:	5.935 cm³
Bohrung:	89 mm
Hub:	79,5 mm
Leistung:	338 kW (460 PS)
	bei 6.500/min

Die Typenbezeichnung des Aston Martin V12 Vanquish verrät es bereits: In diesem Edelbriten schlummert unter der Motorhaube ein mächtiger 12-Zylinder-V-Motor mit 5,9 Litern Hubraum. Er liefert 338 kW (460 PS) bei 6.500 Umdrehungen pro Minute. Das maximale Drehmoment von 542 Nm liegt bei 5.000 Umdrehungen pro Minute an. Pro Zylinder verfügt der Zwölfzylinder über vier Ventile, die über zwei oben liegende Nockenwellen gesteuert werden. Das Drehmoment des vorne eingebauten Motors wird über ein automatisiertes Sechsganggetriebe auf die beiden Hinterräder übertragen. Reizt man das Aggregat voll aus, beschleunigt es den Vanquish in 5,0 Sekunden von 0 auf 100 km/h, und Tempo 160 ist bereits nach 10,5 Sekunden erreicht. Von 2004 an wurde dem Modell, das nun zusätzlich ein „S" in der Typenbezeichnung trug, ein noch stärkerer V12-Motor bei gleichem Hubraum von 5.935 cm³ beschert. Bei 7.000 Umdrehungen pro Minute setzte er 388 kW (528 PS) frei.

The engine

Engine specifications

Type:	3-stroke 60° V engine
Number of cylinders:	12
Displacement:	5935 cm³
Bore:	89 mm
Stroke:	79.5 mm
Power:	338 kW (460 hp)
	at 6500/min

The type designation of the Aston Martin V12 Vanquish already spills the beans: under the hood of this noble Brit slumbers a powerful 12-cylinder V engine with a displacement of 5.9 liters and 338 kW (460 hp) at 6500 rpm. The maximum torque of 542 Nm is reached at 5000 rpm. The engine has four valves per cylinder, which are controlled by overhead camshafts. The torque of the front-mounted engines is transferred to the rear wheels by an automatic six-gear transmission. Exploiting the full capacity of the engine, it is possible to accelerate the Vanquish from 0 to 100 km/h in 5.0 seconds, and a speed of 160 km/h is already reached after 10.5 seconds. Beginning with 2004, the model with an added "S" in the type designation was equipped with a new V12 engine, which provided more power at the same displacement of 5935 cm³. At 7000 rpm, it delivered 388 kW (528 hp).

Der 5,9-Liter-V12 des Aston Martin V12 Vanquish verfügt über DOHC-Vierventil-Technik.

The 5.9-liter V12 engine of the Aston Martin V12 Vanquish uses DOHC four-valve technology.

Der 460 PS starke und 306 km/h schnelle Aston Martin V12 Vanquish verbindet technische Zwölfzylinder-Faszination mit einem überaus edlen britischen Auftritt und formidablen Fahrleistungen.

The Aston Martin Vanquish with 460 hp and a top speed of 306 km/h combines the technical fascination of a 12-cylinder engine with the very classy British appearance and an excellent driving performance.

Die Mercedes-Benz SLK-Klasse ging 2004 umfangreich neu gestaltet in die zweite Generation.
After extensive rework, the second generation of the Mercedes-Benz SLK class was launched in 2004.

Mercedes Benz SLK 350

Fahrzeugdaten

Hersteller:	Mercedes-Benz
Land:	Deutschland
Modell:	SLK 350
Bauzeit:	2004–2011
Länge:	4.082 mm
Breite:	1.788 mm
Höhe:	1.298 mm
Radstand:	2.430 mm
Leergewicht:	1.465 kg
Antriebsart:	Hinterrad
Höchstgeschwindigkeit:	250 km/h (elektronisch begrenzt)
Verbrauch:	10,6 Liter/100 km

Die SLK-Klasse wurde von Mercedes 1996 ins Leben gerufen. Mit ihr bot man einen Roadster an, der kleiner und günstiger als der Mercedes SL war. Die drei Buchstaben werden häufig mit „sportlich", „leicht" und „kompakt" oder „kurz" übersetzt, obwohl „SLK" laut Hersteller keine Abkürzung darstellen soll. Technisch war der SLK ein naher Verwandter der Mercedes C-Klasse. Zu den besonderen Merkmalen des SLK zählte das klappbare Stahldach, das auf Knopfdruck vollständig im Kofferraum verschwindet. 2004 stellte Mercedes-Benz die zweite Generation der SLK-Baureihe vor. Sie trug die interne Bezeichnung „R 171". Der neue Roadster kam bei den Cabrio-Fans sofort gut an. Bis Ende 2009 wurden mehr als 220.000 Einheiten verkauft. Die zweite SLK-Generation war bis 2011 erhältlich, bevor sie von der Baureihe R 172 abgelöst wurde. Seitdem behauptet sich auch diese dritte Generation des dynamischen und schnittig gestalteten Roadsters sehr erfolgreich bei den Fans.

Specifications

Manufacturer:	Mercedes-Benz
Country:	Germany
Model:	SLK 350
Produced:	2004–2011
Length:	4082 mm
Width:	1788 mm
Height:	1298 mm
Wheelbase:	2430 mm
Empty weight:	1465 kg
Type of drive:	Rear-wheel drive
Max. speed:	250 km/h (electronically limited)
Fuel consumption:	10.6 l/100 km

With the SLK class launched in 1996, Mercedes offered a roadster that was smaller and cheaper than the Mercedes SL. The three letters "SLK" were often interpreted as "sporty," "light" and "compact" or "short" ("kompakt" and "kurz" in German); however, according to the manufacturer, they were not meant as abbreviations. Technically, the SLK was a near relative of the Mercedes C class. The special features of the SLK were the foldable steel roof, which could completely disappear in the trunk at the press of a button. In 2004, Mercedes-Benz introduced the second generation of the SLK series. The new roadster with the internal designation "R 171" was immediately successful with cabriolet fans. Until the end of 2009, more than 220,000 units were sold. The second SLK generation was available until 2011, when it was replaced by the R 172 series. Since then, the third generation of this dynamic and sleek roadster has also been very popular with the fans.

Der Motor

Motordaten

Bauart:	Viertakt-90-Grad-V-Motor
Zylinderzahl:	6
Takte:	4
Hubraum:	3.498 cm³
Bohrung:	92,9 mm
Hub:	86 mm
Leistung:	200 kW (272 PS) bei 6.000/min

Für den SLK wurde mit dem M272 eigens ein V-Sechszylinder entwickelt. Außergewöhnlich war der Zylinderbankwinkel von 90 Grad. Er resultierte daraus, dass der M272 vom 90-Grad-V8-Motor M2732 abgeleitet worden war. Daher besaß der M272 eine Ausgleichswelle. Sie war im Zylinder-V angeordnet und sorgte für den Ausgleich freier Massekräfte erster Ordnung. Der Gaswechsel erfolgte über je vier Ventile, die über zwei oben liegende Nockenwellen gesteuert wurden. Die Ventile wurden über Rollenschlepphebel betätigt. Der M272 wurde in verschiedenen Varianten gebaut: Die Palette reichte von der 2,5-Liter-Version mit 150 kW (204 PS) bis zum Sportmotor mit 3,5 Liter Hubraum und 232 kW (316 PS). Im SLK 350 kam der M272 E 35 mit 3,5 Litern und 200 kW (272 PS) zum Einsatz. Sein maximales Drehmoment von 350 Nm stellte er von 2.400 bis 5.000 Touren bereit. Je nach Ausstattung wurde das Drehmoment über ein Sechsgang-Schaltgetriebe oder eine Siebengang-Automatik übertragen.

The engine

Engine specifications

Type:	Four-stroke 90° V engine
Number of cylinders:	6
Displacement:	3498 cm³
Bore:	92.9 mm
Stroke:	86 mm
Power:	200 kW (272 hp) at 6000 rpm

The M272 was a V6 engine specifically developed for the SLK. The bank angle of 90° was unusual and resulted from the fact that the M272 was derived from the 90° V8 engine M2732. For this reason, the M272 had a balance shaft in the "V", which balanced free inertial forces of the first order. Gas exchange took place via four valves, which were controlled by two overhead camshafts and actuated by roller cam followers. Several versions of the M272 were built. The range extended from the 2.5-liter version with 150 kW (204 hp) to the sports motor with a displacement of 3.5 liters and 232 kW (316 hp). The SLK 350 used the M272 E 35 with 3.5 liters and 200 kW (272 hp), which delivered its maximum torque of 350 Nm between 2400 and 5000 rpm. Depending on configuration, the torque was transferred by a manual six-gear transmission or an automatic seven-gear transmission.

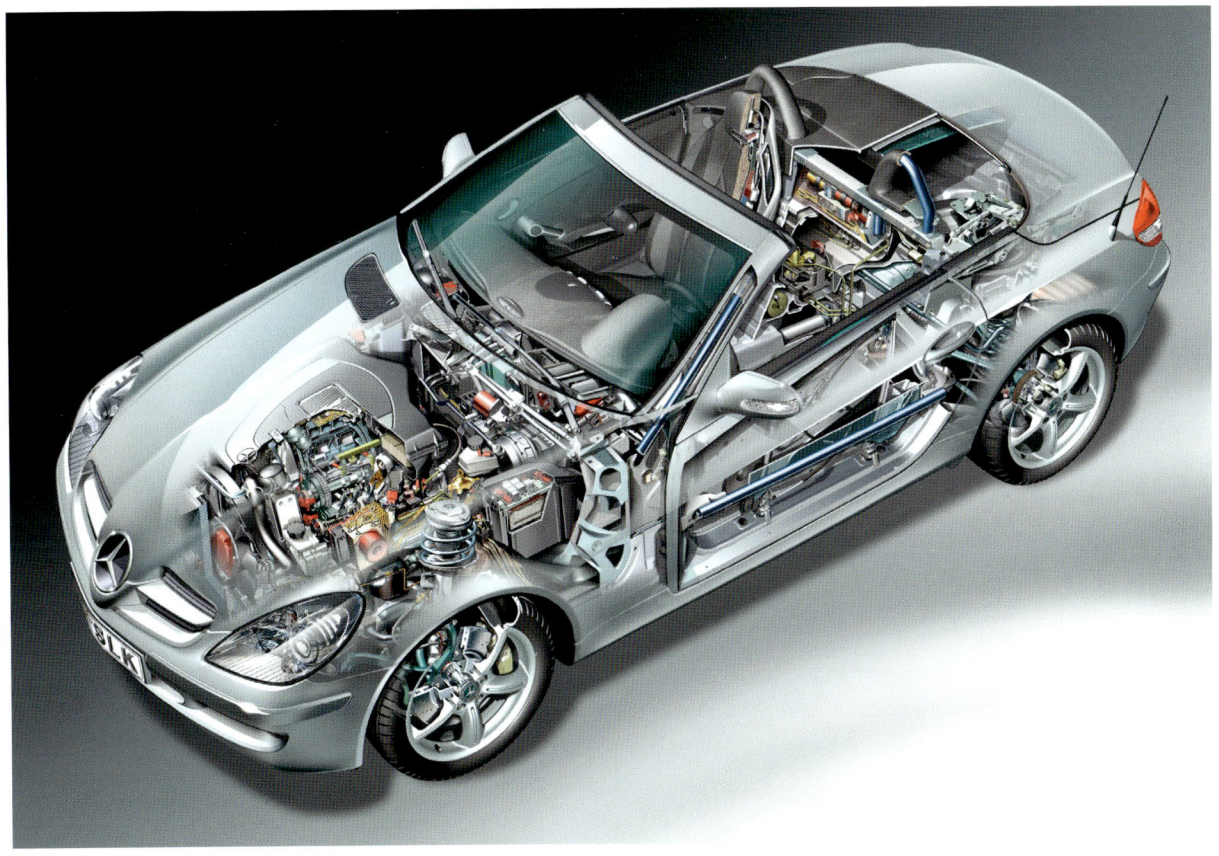

Mit dem SLK 350 ging 2004 auch ein neuer V6-Motor an den Start, der 200 kW (272 PS) leistete.

Together with the SLK 350, a new V6 engine was launched in 2004, providing 200 kW (272 hp).

Mit dem SLK gelang Mercedes ein richtig großer Wurf im Hinblick auf viel Fahrspaß und hohe Fahrdynamik in einem potenten, handlichen, aber dennoch alltags- und praxistauglichen Roadster.

With the SLK, Mercedes hit the jackpot with regard to driving pleasure, driving dynamics and a powerful, handy and yet practical roadster.

Der BMW Hydrogen 7 aus dem Jahr 2006 wurde von einem Wasserstoff-Verbrennungsmotor angetrieben.

The BMW Hydrogen 7 built in 2006 was driven by a hydrogen combustion engine.

BMW Hydrogen 7

Fahrzeugdaten

Hersteller:	BMW
Land:	Deutschland
Modell:	Hydrogen 7
Bauzeit:	2006
Länge:	5.179 mm
Breite:	1.902 mm
Höhe:	1.489 mm
Radstand:	3.128 mm
Leergewicht:	2.385 kg
Antriebsart:	Hinterrad
Höchstgeschwindigkeit:	230 km/h
Verbrauch (Benzin):	13,9 Liter/100 km
Verbrauch Wasserstoff:	3,6 kg/100 km (entspricht 13,3 Liter Benzin)

Unter der Bezeichnung Hydrogen 7 präsentierte BMW im Jahr 2006 eine Luxus-Limousine, die von einem Wasserstoff-Verbrennungsmotor angetrieben wurde. Auf Basis des Modells 760Li entstanden im BMW-Werk Dingolfing insgesamt 100 Exemplare des Hydrogen 7. Das Auto wurde von BMW nicht zum Kauf angeboten, sondern konnte von Kunden lediglich geleast werden. In den USA wurde der BMW einem ausgewählten Nutzerkreis zur Verfügung gestellt. Zu den prominenten Fahrern gehörte unter anderem Plácido Domingo. Der Motor des Hydrogen 7 basierte auf dem Antriebsaggregat des 760i, war aber für den Einsatz von Wasserstoff modifiziert worden. Beim Betrieb mit diesem Treibstoff erreichte der BMW eine Reichweite von mehr als 200 Kilometern. Das Fahrzeug konnte aber auch im herkömmlichen Benzin-Modus gefahren werden – und so weitere 500 Kilometer zurücklegen.

In einem Feldversuch waren die Fahrzeuge bis Ende 2009 im Einsatz. Danach wurde das Projekt von BMW nicht weiter fortgesetzt.

Specifications

Manufacturer:	BMW
Country:	Germany
Model:	Hydrogen 7
Produced:	2006
Length:	5179 mm
Width:	1902 mm
Height:	1489 mm
Wheelbase:	3128 mm
Empty weight:	2385 kg
Type of drive:	Rear-wheel drive
Max. speed:	230 km/h
Fuel consumption (gasoline):	13.9 l/100 km
Fuel consumption (hydrogen):	3.6 kg/100 km (corresponding to 13.3 liters of gasoline)

In 2006, BMW introduced the Hydrogen 7, a limousine driven by a hydrogen combustion engine. Based on the model 760Li, 100 units of this car were built in the BMW factory at Dingolfing. It was not sold by BMW but could only be leased. In the USA, it was provided for an exclusive circle of users, among others Plácido Domingo. The engine of the Hydrogen 7 was based on the engine of the 760i, albeit modified for the use of hydrogen. With this fuel, the car achieved a range of 200 km but it could also be driven in traditional gasoline mode and thus travel another 500 km.

The cars were used in a field test until the end of 2009. After that, BMW discontinued the project.

Der Motor

Motordaten

Bauart:	Viertakt-60-Grad-V-Motor
Zylinderzahl:	12
Hubraum:	5.972 cm³
Bohrung:	89 mm
Hub:	80 mm
Leistung:	191 kW (260 PS)
	bei 5.100/min

Der V12-Motor der Wasserstoff-Luxuslimousine wurde vom Benzin-Triebwerk des BMW 760i abgeleitet. Während der Originalmotor für eine Leistung von 326 kW (445 PS) sorgte, erreichte das modifizierte Aggregat im Hydrogen 7 nur 191 kW (260 PS) bei 5.100/min. Bei 4.300 Touren erreichte er mit 390 Nm sein maximales Drehmoment. Immerhin ließ sich der Wagen damit in 9,5 Sekunden von 0 auf 100 km/h beschleunigen. Die Höchstgeschwindigkeit war auf 230 km/h begrenzt. Im Benzin-Betrieb arbeitete der Motor des Hydrogen 7 als Direkteinspritzer. Im Wasserstoff-Betrieb erfolgte die Gemischbildung dagegen bereits in den Ansaugkanälen. Der Wasserstofftank des BMW nahm rund 8 Kilogramm Flüssigwasserstoff auf, was etwa 170 Litern entsprach. Der Benzintank umfasste dagegen ein Volumen von 74 Litern. Das bivalente Motoren-konzept des Hydrogen 7 war im Sinne der Alltagstauglichkeit des Wagens unentbehrlich, da das Netz von Wasserstoff-Tankstellen viel zu klein war.

The engine

Engine specifications

Type:	Four-stroke 60° V engine
Number of cylinders:	12
Displacement:	5972 cm³
Bore:	89 mm
Stroke:	80 mm
Power:	191 kW (260 hp)
	at 5100 rpm

The V12 engine of the hydrogen limousine was derived from the gasoline engine of the BMW 760i. While the original engine delivered 326 kW (445 hp), the modified engine of the Hydrogen 7 only provided 191 kW (260 hp) at 5100 rpm. At 4300 rpm, it reached its maximum torque of 390 Nm. At least, the car accelerated from 0 to 100 km/h in 9.5 seconds. The top speed was limited to 230 km/h. In gasoline mode, the engine of the Hydrogen 7 used direct injection. In hydrogen mode, the mixture was already generated in the intake ports. The hydrogen tank of the car could accommodate approx. 8 kg liquid hydrogen, which equalled approx. 170 liters, while the gasoline tank had a capacity of 74 liters. The double engine concept of the Hydrogen 7 was essential for practical use as the network of hydrogen stations was much too small.

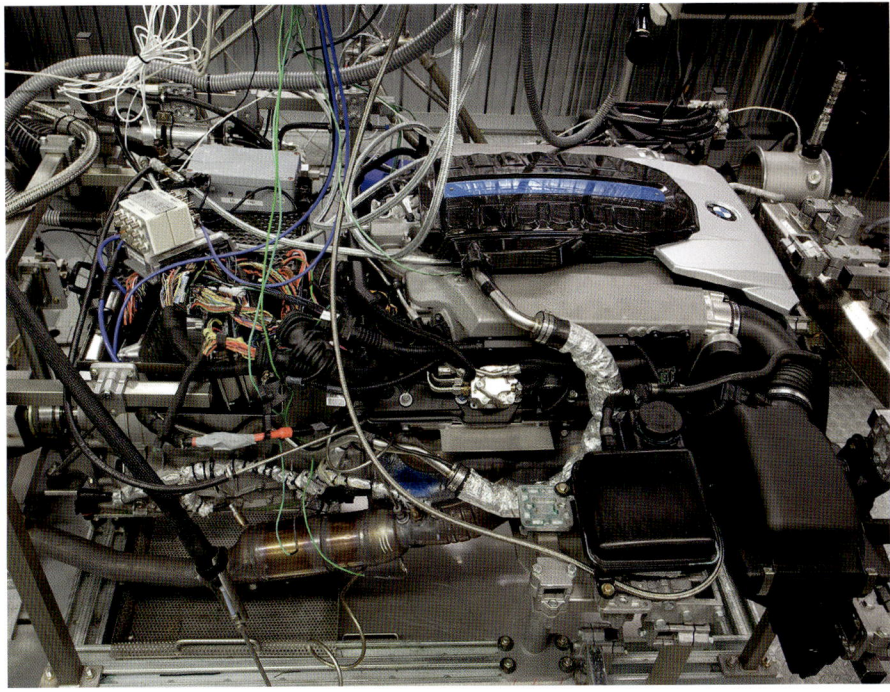

Der V12-Wasserstoff-Verbrennungsmotor des BMW Hydrogen 7 erreichte eine Höchstleistung von 191 kW (260 PS).

The V12 hydrogen combustion engine of the BMW Hydrogen 7 achieved a maximum power of 191 kW (260 hp).

Bevor BMW den Hydrogen 7 in Kleinserie baute, wurde der Wasserstoff-Verbrennungsmotor intensiv auf dem Prüfstand getestet.

Before building a small series of the Hydrogen 7, BMW tested the combustion engine thoroughly.

Auf Basis der Luxus-Limousine 760Li entwickelte BMW das Modell Hydrogen 7, das mit einem Wasserstoff-Verbrennungsmotor ausgestattet war. 2006 produzierte die bayerische Marke insgesamt 100 dieser Fahrzeuge.

BMW developed the Hydrogen 7 based on the limousine 760Li. In 2006, the Bavarian manufacturer build 100 of these vehicles with hydrogen combustion engine.

Der neue Fiat 500, hier in der Version TwinAir von 2010, wird seit 2007 mit großem Erfolg gebaut.
The new Fiat 500 (here, the 2010 TwinAir version) is built successfully since 2007.

Fiat 500 TwinAir

Fahrzeugdaten

Hersteller:	Fiat
Land:	Italien
Modell:	500 TwinAir
Bauzeit:	seit 2007
Länge:	3.546 mm
Breite:	1.627 mm
Höhe:	1.488 mm
Radstand:	2.300 mm
Leergewicht:	1.005 kg
Antriebsart:	Vorderrad
Höchstgeschwindigkeit:	173 km/h
Verbrauch:	4,1 Liter/100 km

Kaum ein anderes Auto kann auf eine längere Tradition zurückblicken als der Fiat 500. Sie begann mit dem Fiat 500 Topolino, der zwischen 1936 und 1948 die Produktionshallen verließ. Seine erste Wiederauferstehung feierte der kleine Italiener als Nuova 500 im Jahr 1957. Er prägte das Straßenbild nachhaltig – was bei rund 3,4 Millionen bis 1975 verkauften Einheiten nicht verwundert. 2007 war die Zeit wieder reif für den Fiat 500. Er ist größer, stärker und schneller als sein Vorgänger aus den 1950er Jahren, der neue 500er versprüht aber dennoch den unverwechselbaren Charme. Bei dem neuen Viersitzer wird das Thema Sicherheit groß geschrieben: Er ist beispielsweise mit sechs Airbags sowie einem Knieairbag für den Fahrer ausgestattet. Beim Euro-NCAP-Crashtest war der neue Fiat 500 der erste Kleinwagen, der die Höchstwertung erreichte.

Um allen Wünschen gerecht zu werden, bietet Fiat den neuen 500 in zahlreichen Modellvarianten und mit Motoren mit zwei und vier Zylindern an. Im sportlichen Abarth 695 sind sogar 230 km/h Höchstgeschwindigkeit möglich.

Specifications

Manufacturer:	Fiat
Country:	Italy
Model:	500 TwinAir
Produced:	since 2007
Length:	3546 mm
Width:	1627 mm
Height:	1488 mm
Wheelbase:	2300 mm
Empty weight:	1005 kg
Type of drive:	Front-wheel drive
Max. speed:	173 km/h
Fuel consumption:	4.1 l/100 km

There is nearly no other car with a longer tradition than the Fiat 500. Its history began with the Fiat 500 Topolino built from 1936 to 1948. The small Italian enjoyed its first resurrection as Nuova 500 in 1957. It became a typical element of the streetscape; not surprisingly, as approx. 3.4 million units were sold until 1975. In 2007, the time was again ripe for the Fiat 500. It was now larger, more powerful and faster than its predecessor from the 1950s, but it still aired its distinctive charme. The new four-seater provides safety with a capital S! For instance, it has six airbags plus a knee airbag for the driver. The Fiat 500 was the first subcompact car to achieve the highest rating in the Euro NCAP crash test.

In order to fulfill all requirements, the new Fiat 500 is available in numerous versions with 2- and 4-cylinder engines. The Abarth 695 can even reach a top speed of 230 km/h.

Der Motor

Motordaten

Bauart:	Viertakt-Reihenmotor mit Turbolader
Zylinderzahl:	2
Hubraum:	875 cm³
Bohrung:	80,5 mm
Hub:	86 mm
Leistung:	63 kW (85 PS) bei 5.500/min

„TwinAir" lautet die Bezeichnung des seit 2010 im Fiat 500 eingebauten Zweizylinder-Reihenmotors. Es handelt sich um einen Parallel-Twin-Motor, der sich durch seinen prinzipiell einfachen und damit kostengünstigen Aufbau auszeichnet. Für die Zweizylinder-Konstruktion des Parallel-Twin-Aggregats spricht der im Vergleich zu Verbrennungsmotoren mit mehr Zylindern höhere Wirkungsgrad. Beide Zylinder des mit einem Turbolader versehenen TwinAir sind mit je zwei Ein- und zwei Auslassventilen ausgestattet, die über eine gemeinsame oben liegende und über eine Steuerkette angetriebene Nockenwelle angesteuert werden. Das Verdichtungsverhältnis der wassergekühlten Maschine liegt bei 10:1. Ihre maximale Leistung von 63 kW (85 PS) stellt sie bei 5.500 Umdrehungen pro Minute zur Verfügung. Ihr höchstes Drehmoment von 145 Nm entwickelt sie bereits bei 1.900 Touren. Der Motor wird als Turbo-Triebwerk sowie als Saugmotor gebaut. Außer im Fiat 500 findet man das TwinAir-Aggregat auch in anderen Modellen.

The engine

Engine specifications

Type:	Four-stroke in-line engine with turbo charger
Number of cylinders:	2
Displacement:	875 cm³
Bore:	80.5 mm
Stroke:	86 mm
Power:	63 kW (85 hp) at 5500 rpm

The 2-cylinder in-line engine used in the Fiat 500 since 2010 is called "TwinAir." It is a parallel twin engine which a simple and thus cost-effective design. One reason for using a 2-cylinder design was the higher degree of efficiency in comparison with combustion engines with more cylinders. Both cylinders of the turbo-charged Twinair have two intake and two outlet valves each which are controlled by a common overhead camshaft, which in turn is driven by a timing chain. The compression ratio of the water-cooled engine amounts to 10:1. The maximum power of 63 kW (85 hp) is reached at 5500 rpm, and the maximum torque of 145 Nm at 1900 rpm. The engine is built as turbo engine and as suction engine. The TwinAir engine is also used in other models than the Fiat 500.

Die Zweizylinder-Konstruktion des TwinAir von 2010 gehorcht vor allem Downsizing-Prinzipien.

The 2-cylinder design of the 2010 TwinAir particularly follows downsizing concepts.

Mit dem neuen Fiat 500 gelang dem italienischen Hersteller 2007 nicht zuletzt dank der Retro-Welle das unglaubliche Comeback eines schon totgesagten Kleinwagens.

With the introduction of the new Fiat 500 in 2007, the Italian manufacturer succeeded with the incredible comeback of the subcompact car already declared dead – not least because of the retro trend.

Der Toyota Prius II besaß noch einen 1,5-Liter-Verbrennungsmotor mit 56 kW (77 PS).
The Toyota Prius II still had a 1.5-liter combustion engine with 56 kW (77 hp).

Toyota Prius III

Fahrzeugdaten

Hersteller:	Toyota
Land:	Japan
Modell:	Prius III
Bauzeit:	2009–2016
Länge:	4.480 mm
Breite:	1.745 mm
Höhe:	1.490 mm
Radstand:	2.700 mm
Leergewicht:	1.370–1.420 kg
Antriebsart:	Vorderrad
Höchstgeschwindigkeit:	180 km/h
Verbrauch:	3,9 Liter/100 km

Mit dem Prius III brachte Toyota 2009 die dritte Auflage des seit 1997 gebauten Wagens auf den Markt: ein Auto, das auf ein intelligentes Zusammenspiel von Elektro- und Benzinantrieb setzte, bis 45 km/h rein elektrisch fahren konnte und zudem lediglich 3,9 Liter Super pro 100 Kilometer verbrauchen sollte; nicht zu vergessen der geringe CO_2-Ausstoß von 89 g/100 km. Doch nicht nur diese Eckdaten sowie die Systemleistung von 100 kW (136 PS), bestehend aus 73 kW (99 PS) des Benzinmotors und der zusätzlichen Leistung der E-Maschine, machten die dritte Auflage des Hybrid-Automobils attraktiv. Auch der Preis von 24.950 Euro rückte diese Zukunftstechnologie für viele Kunden in erreichbare Dimensionen. Zukunftsweisend war aber nicht nur das Antriebs-konzept. Auch das Interieur mit futuristisch geschwungenem „schwebendem" Armaturenbrett mit Flachbildschirm signalisierte die Fortschrittlichkeit dieses Wagens. Für 1.000 Euro Aufpreis gab es zudem ein Solardach zur Stromversorgung der Klimaanlage.

Specifications

Manufacturer:	Toyota
Country:	Japan
Model:	Prius III
Produced:	2009–2016
Length:	4480 mm
Width:	1745 mm
Height:	1490 mm
Wheelbase:	2700 mm
Empty weight:	1370–1420 kg
Type of drive:	Front-wheel drive
Max. speed:	180 km/h
Fuel consumption:	3.9 l/100 km

With the Prius III introduced in 2009, Toyota launched the third edition of a car that was built since 1997. This was a car that relied on the intelligent combination of electrical drive and gasoline engine, could reach up to 45 km/h purely with the electrical drive and consumed only 3.9 liters premium gasoline per 100 km. Besides, it only emitted 89 g/100 km CO_2. But these specifications and the system power of 100 kW (136 hp) comprised of 73 kW (99 hp) of the gasoline engine and the additional power of the electromotor were not the only reasons for the attractiveness of this third edition of the hybrid automobile. The price of 24,950 Euros made this future technology available for many customers. But the drive concept was not the only visionary aspect. The interior with its futuristically curved head-up display and a flat screen also proved the progressiveness of the car. For an added 1000 Euros, you could also get a solar roof as power supply for the air conditioning.

Der Motor

Motordaten

Bauart:	Viertakt-Reihenmotor kombiniert mit Elektromotor
Zylinderzahl:	4
Hubraum:	1.797 cm³
Bohrung:	80,5 mm
Hub:	88,3 mm
Leistung:	73 kW (99 PS) bei 5.200/min, Systemleistung 100 kW (136 PS)

Der Benzinmotor war das primäre Antriebsaggregat des Toyota Prius III. Er brachte es bei 5.200 Umdrehungen pro Minute auf eine Leistung von 73 kW (99 PS). In jedem seiner vier Zylinder kümmerten sich je zwei Ein- und Auslassventile um einen effizienten Gaswechsel. Die Steuerung der Ventile erfolgte über zwei oben liegende Nockenwellen. Das maximale Drehmoment von 142 Nm lag bei 4.000 Touren an. An den Verbrennungsmotor war eine E-Maschine – eine Kombination aus Elektromotor und Generator – gekoppelt. Zusammen mit dem Verbrennungsmotor ergab sich so eine Gesamtsystemleistung von 100 kW (136 PS). Beeindruckend war das maximale Drehmoment des Elektromotors von 207 Nm, das konstruktionsbedingt sofort und nicht erst bei einer gewissen Drehzahl anlag. Die Speicherung elektrischer Energie erfolgte über Nickel-Metallhydrid-Batterien. Beide Aggregate konnten im Fahrbetrieb auch unabhängig voneinander genutzt werden, und es standen verschiedene Fahrmodi zur Verfügung.

The engine

Engine specifications

Type:	Four-stroke in-line engine in combination with an electromotor.
Number of cylinders:	4
Displacement:	1797 cm³
Bore:	80.5 mm
Stroke:	88.3 mm
Power:	73 kW (99 hp) at 5200/min, total system power 100 kW (136 hp)

The gasoline engine was the primary drive of the Toyota Prius III. At 5200 rpm, it provided 73 kW (99 hp). At each of the four cylinders, two intake valves and two outtake valves arranged for an efficient gas exchange. The valves were controlled by two overhead camshafts. The maximum torque of 142 Nm was reached at 4000 rpm. Coupled with the combustion engine was the e-drive, a combination of electromotor and generator. Together with the combustion engine, it produced a total system power of 100 kW (136 hp). By design, the impressive maximum torque of the electromotor of 207 Nm was immediately available and not only at a certain engine speed. Electrical energy was stored in nickel-metal hydride batteries. Both engines could be used independently of each other. Various driving modes were available.

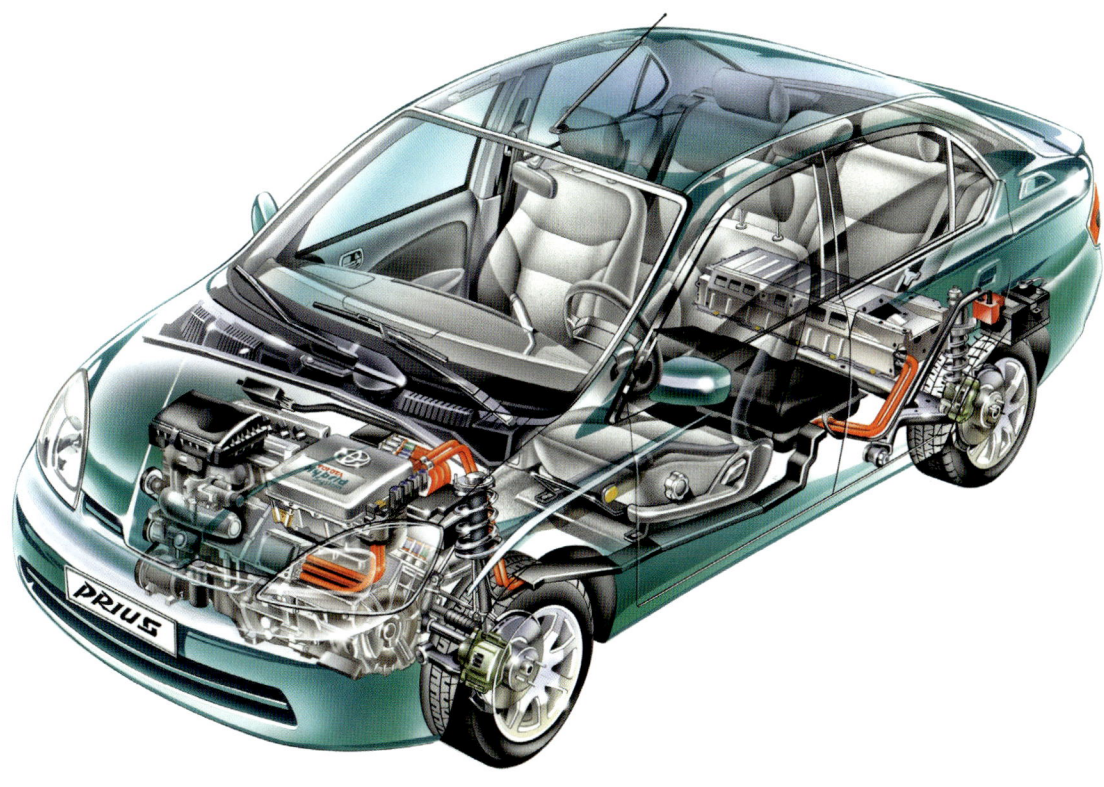

Grafik des hybriden Antriebskonzepts des Toyota Prius I, der 2000 auf den deutschen Markt kam.

Diagram of the hybrid drive concept of the Toyota Prius I, which was launched on the German market in 2000.

An den 1,8-Liter-Benzinmotor mit 73 kW (99 PS) ist ein Elektromotor gekoppelt.

The 1.8-liter gasoline engine with 73 kW (99 hp) is coupled with an electromotor.

Schnittmodell des Toyota Prius III Plug-in-Hybrid aus dem Jahr 2010.

Cutaway of the 2010 Toyota Prius III Plug-In Hybrid.

Mit dem Prius III setzte Toyota auf das intelligente Zusammenspiel von Elektro- und Verbrennerantrieb und machte die Hybrid-Technologie salonfähig.

With the Prius III, Toyota relied on the intelligent combination of electromotor and combustion engine and thus made hybrid drive technology socially acceptable.

Der Ferrari 458 Spider war ein atemberaubender Supersportler, der auch optisch beeindruckte.

The Ferrari 458 Spider was a breathtaking super sports car with an impressive visual appearance.

Ferrari 458 Spider

2011

Fahrzeugdaten

Hersteller:	Ferrari
Land:	Italien
Modell:	458 Spider
Bauzeit:	2011–2015
Länge:	4.527 mm
Breite:	1.937 mm
Höhe:	1.211 mm
Radstand:	2.650 mm
Leergewicht:	1.535 kg
Antriebsart:	Hinterrad
Höchstgeschwindigkeit:	320 km/h
Verbrauch:	11,8 Liter/100 km

Keine andere Marke aus der Welt der Sportwagen weckt so viele Emotionen wie Ferrari. Für zahllose Fans auf der ganzen Welt steht der Name für Leidenschaft, Kraft, Eleganz, Exklusivität und Sportlichkeit. Firmengründer Enzo Ferrari war schon zu Lebzeiten eine legendäre Gestalt – und die roten Rennwagen aus Maranello feierten in der Formel 1 ab 1950 zahllose Siege und WM-Titel.

Der Nimbus des sportlichen Erfolgs übertrug sich auf die Straßensportwagen aus dem Hause Ferrari. 2011 präsentierte die italienische Marke mit dem 458 Spider eine offene Variante des zwei Jahre zuvor vorgestellten Ferrari 458 Italia. Der Spider besaß einen V8-Zylinder-Mittelmotor mit 4,5 Litern Hubraum, der 418 kW (570 PS) bei 9000/min leistete. Das reichte, diesem Sportwagen zu atemberaubenden Fahrleistungen zu verhelfen: Von 0 auf 100 km/h beschleunigte der 458 Spider in nur 3,4 Sekunden. Und erst bei Tempo 320 erreichte der Supersportler seine Höchstgeschwindigkeit.

Specifications

Manufacturer:	Ferrari
Country:	Italy
Model:	458 Spider
Produced:	2011–2015
Length:	4527 mm
Width:	1937 mm
Height:	1211 mm
Wheelbase:	2650 mm
Empty weight:	1535 kg
Type of drive:	Rear-wheel drive
Max. speed:	320 km/h
Fuel consumption:	11.8 l/100 km

No other brand name in the world of sports cars is as highly emotional as Ferrari. For numerous fans all around the world, the name evokes a sense of passion, power, elegance, exclusiveness and sportiveness. Company founder Enzo Ferrari had already been a legend in his life times, and beginning with 1950, the red racing cars from Maranello could celebrate many Formula 1 wins and world championship titles. The nimbus of sports success was also transferred to the street sports cars by Ferrari. In 2011, the Italian brand presented the 458 Spider, which was an open version of the Ferrari 458 Italia introduced two years ago. The Spider had a V8 midship engine with a displacement of 4.5 liters and an engine power of 418 kW (570 hp) at 9000 rpm. This was sufficient to grant this sports car some breathtaking driving characteristics. Accelerating from 0 to 100 km/h took the 458 Spider just 3.4 seconds, and the speed limit was reached only at 320 km/h.

Der Motor

Motordaten

Bauart:	Viertakt-90-Grad-V-Motor
Zylinderzahl:	8
Hubraum:	4.499 cm³
Bohrung:	94 mm
Hub:	81 mm
Leistung:	418 kW (570 PS)
	bei 9.000/min

Der Ferrari 458 Spider besaß einen V8-Motor mit Benzindirekteinspritzung, Trockensumpf-schmierung und einem Hubraum von 4.499 cm³. Der Zylinderwinkel betrug 90 Grad. Das Sportwagen-Triebwerk brachte es auf eine Höchstleistung von 418 kW (570 PS) bei 9.000/min. Eine so hohe Drehzahl war von Motoren dieser Hubraumgröße im Serienbereich zuvor nicht erreicht worden. Auch die spezifische Leistung von 93 kW/Liter war ein neuer Rekord für einen Straßensportwagen mit Saugmotor. Das Ferrari-Aggregat verfügte über vier Ventile pro Zylinder sowie eine variable Ventilsteuerung. Das maximale Drehmoment des V8 lag bei 540 Nm bei 6.000 Umdrehungen pro Minute. Das Verdichtungsverhältnis betrug 12,5:1. Im Jahr 2011 wurde diese Ferrari-Konstruktion von einer Fachjury zum „Motor des Jahres" gewählt.

Im Bereich der Kraftübertragung setzten die Ingenieure auf ein Siebengang-Doppelkupplungs-getriebe, das für blitzschnelle Gangwechsel ohne Drehmomentunterbrechungen sorgte.

The engine

Engine specifications

Type:	Four-stroke 90° V engine
Number of cylinders:	8
Displacement:	4499 cm³
Bore:	94 mm
Stroke:	81 mm
Power:	418 kW (570 hp)
	at 9000 rpm

The Ferrari 458 Spider had a V8 engine with direct gasoline injection, dry sump lubrication and a displacement of 4499 cm³. The cylinders were arranged at an angle of 90°. The maximum power of 418 kW (570 hp) was reached at 9000 rpm. Never before had a mass-produced engine of this displacement achieved such a high engine speed. The specific power of 93 kW/l was also a new record for a street sports car with suction engine. The Ferrari engine had four valves per cylinder and variable valve control. The maximum torque of the V8 amounted to 540 Nm at 6000 rpm. The compression ratio was 12.5:1. In 2011, this Ferrari design was awarded "engine of the year" by a specialist jury.

For the power train, the engineers relied on a 7-gear twin clutch transmission, which allowed rapid gear changes without torque interruptions.

Das Ferrari-V8-Triebwerk wurde 2011 als „Motor des Jahres" ausgezeichnet.
The Ferrari V8 was awarded "engine of the year" in 2011.

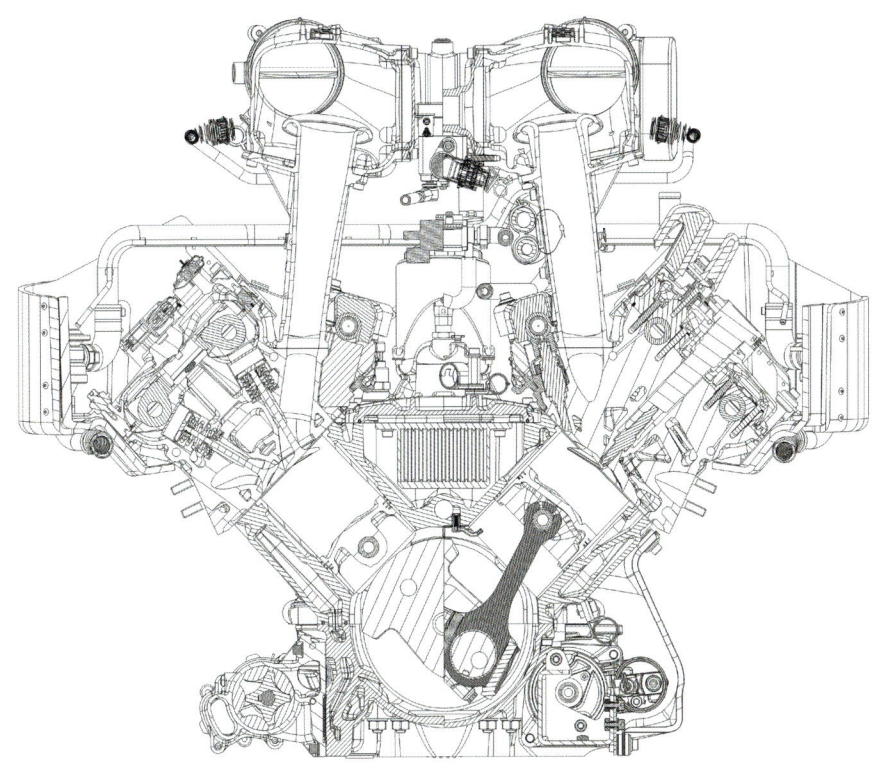

Im Antriebsaggregat des Ferrari 458 kam Technologie aus dem Motorsport zum Einsatz.
Motorsports technology was used in the engine of the Ferrari 458.

Im Ferrari 458 Spider sorgte ein Mittelmotor für Vortrieb – und der 4,5-Liter-V8 überzeugte durch Leistung und Drehfreudigkeit. Das Antriebsaggregat des Italieners leistete 418 kW (570 PS) bei 9.000/min.

The Ferrari 458 Spider was driven by a midship engine. The 4.5-liter V8 engine boasted a rev-happy performance. The engine delivered 418 kW (570 hp) at 9000 rpm.

Der VW Up! musste seine Qualitäten selbstredend auch im Windkanal beweisen.
Of course, the VW Up! had to prove its worth in the wind tunnel.

VW Up!

2011

Fahrzeugdaten

Hersteller:	Volkswagen
Land:	Deutschland
Modell:	Up!
Bauzeit:	seit 2011
Länge:	3.540 mm
Breite:	1.641 mm
Höhe:	1.489 mm
Radstand:	2.420 mm
Leergewicht:	929 kg
Antriebsart:	Vorderrad
Höchstgeschwindigkeit:	171 km/h
Verbrauch:	5,5 Liter/100 km

Volkswagen präsentierte mit dem gerade einmal 3,54 Metern langen Up! Ende 2011 den Nachfolger des VW Fox. Er ist das kleinste Modell der VW-Produktpalette. Er ist zwar primär für den Stadtverkehr gedacht, vermag aber auch auf längeren Strecken zu überzeugen. Um die Produktionskosten zu senken, wird der Wagen nach dem Baukastensystem bei VW in Bratislava, Slowakei, gefertigt. Für entfernter liegende Märkte erfolgt die Montage auch in Indien und Russland. Neben den Varianten mit Benzinmotoren gibt es den Up! seit November 2012 auch mit Erdgas-Antrieb. In ihm arbeitet ein 50 kW (68 PS) starker Dreizylinder-Otto-Reihenmotor. Mit 16,3 Sekunden von 0 auf 100 km/h beschleunigt er geringfügig langsamer als mit Benzinmotor. Seit 2013 gibt es den Up! auch als Elektromobil mit einer permanent angeregten Synchronmaschine. Sie leistet 60 kW (82 PS), die zwischen 2.800 und 12.000 Umdrehungen pro Minute abgeben. Seine Höchstgeschwindigkeit beträgt 130 km/h.

Specifications

Manufacturer:	Volkswagen
Country:	Germany
Model:	Up!
Produced:	since 2011
Length:	3540 mm
Width:	1641 mm
Height:	1489 mm
Wheelbase:	2420 mm
Empty weight:	929 kg
Type of drive:	Front-wheel drive
Max. speed:	171 km/h
Fuel consumption:	5.5 l/100 km

In the end of 2011, Volkswagen presented the Up! as the successor of the VW Fox. With a length of merely 3.54 m, this is the smallest model in the VW portfolio. It is primarily designed for local city transport, but can also successfully handle longer distances. To lower the production costs, the car is built modularly in the VW factory of Bratislava, Slovakia. For markets further away, assembly is also done in India and Russia. In addition to versions with gasoline engines, the Up! is also available with a natural-gas drive. This version uses a 3-cylinder in-line Otto engine with 50 kW (68 hp). Going from 0 to 100 km/h in 16.3 seconds, the acceleration is only slightly lower than with the gasoline engine. Since 2013, the Up! is also available as an electric car with a permanently excited synchronous motor. It deliveres a power of 60 kW (82 hp) in the range of 2800 to 12.000 rpm. The top speed amounts to 130 km/h.

Der Motor

Motordaten

Bauart:	Viertakt-Reihenmotor
Zylinderzahl:	3
Hubraum:	999 cm³
Bohrung:	74,5 mm
Hub:	76,4 mm
Leistung:	55 kW (75 PS)
	bei 6.200/min

Im VW Up! arbeitet ein vergleichsweise kleiner Dreizylinder-Motor mit 999 cm³ Hubraum. Er trägt die Bezeichnung 1.0 MPI und wird auch in anderen Einstiegsmodellen des Volkswagen-Konzerns, wie etwa dem Polo, verbaut. Damit wirkt der Konzern dem Trend entgegen, dass selbst in sehr kleinen Autos immer größere und leistungsstärkere Motoren eingebaut werden. 2011 wurde der Dreizylinder-Reihenmotor zunächst mit Multi-Point-Einspritzung mit 44 kW (60 PS) und 55 kW (75 PS) angeboten. Seit Mitte 2016 ist er wahlweise auch mit Direkteinspritzung und Turboaufladung sowie einer gesteigerten Leistung von 66 kW (90 PS) verfügbar. Die maximale Leistung gibt der 55-kW-Motor bei 6.200 Umdrehungen pro Minute ab. Sein höchstes Drehmoment von 95 Nm entfaltet er bei 3.000 bis 4.300 Touren. Als Sonderausstattung steht die BlueMotion-Technologie zur Wahl. In dieser Version zeichnet sich der VW Up! durch einen noch geringeren Verbrauch von knapp über 4 Litern/100 Kilometer aus.

The engine

Engine specifications

Type:	Four-stroke in-line engine
Number of cylinders:	3
Displacement:	999 cm³
Bore:	74.5 mm
Stroke:	76.4 mm
Power:	55 kW (75 hp)
	at 6200 rpm

The VW Up! has a relatively small 3-cylinder engine with a displacement of 999 cm³. Designated as 1.0 MPI, it can also be found in other entry-level cars by Volkswagen, e.g. the Polo. This decision was made against the trend of using ever larger and more powerful engines in very small cars. In 2011, the 3-cylinder in-line motor was available with multipoint injection and a power rating of 44 kW (60 hp) or 55 kW (75 hp), and since 2016 it can also be purchased optionally with direct injection, turbo charging and an increased power of 66 kW (90 hp). The maximum power of the 55 kW engine is reached at 6200 rpm, the maximum torque of 95 Nm between 3000 and 4300 rpm. BlueMotion technology is available optionally. In the respective version, the VW Up! is distinguished by an even lower fuel consumption of slightly more than 4 l/100 km.

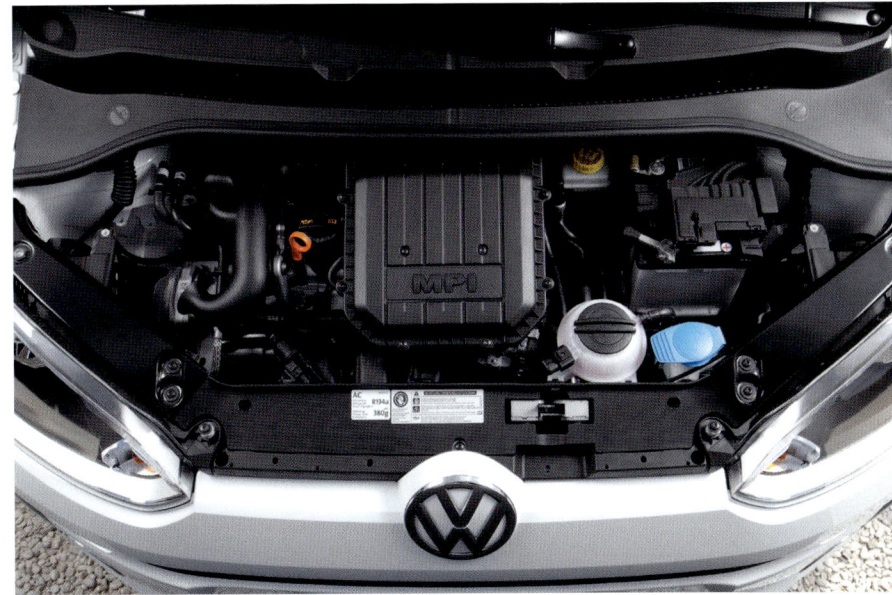

Den Dreizylinder-Benzinmotor 1.0 MPI gibt es mit 44 kW (60 PS) und 55 kW (75 PS).
The 3-cylinder gasoline engine 1.0 MPI is available with 44 kW (60 hp) and 55 kW (75 hp).

Der VW Up! setzte Ende 2011 die erfolgreichen Karrieren seiner Volkswagen-Vorgänger Fox und Lupo fort. Als Antriebsalternative steht bei ihm auch ein Erdgasantrieb zur Verfügung.

Since the end of 2011, the VW Up! continues the successful careers of its predecessors Fox und Lupo. It is also available with an alternative natural-gas drive.

Der Audi A6 3.0 TDI (hier Bj. 2014) verbindet hohe Effizienz mit toller Performance.
The Audi A6 3.0 TDI (2014) combines high efficiency and great performance.

Audi A6 3.0 TDI V6

2011

Fahrzeugdaten

Hersteller:	Audi
Land:	Deutschland
Modell:	A6 3.0 TDI V6
Modell:	Limousine
Baujahr:	seit 2011
Länge:	4.933 mm
Breite:	1.874 mm
Höhe:	1.455 mm
Radstand:	2.912 mm
Leergewicht:	1.845 kg
Antriebsart:	Allrad
Höchstgeschwindigkeit:	250 km/h (abgeriegelt)
Verbrauch:	5,1 Liter/100 km

*Daten für Audi A6 3.0 TDI V6 2014

Zahlreiche Autos werden heute nach modularen Systemen zusammengebaut, was insbesondere Entwicklungs- wie auch Produktionskosten einspart. Dies trifft auch auf den Audi A6 C7 zu, seines Zeichens bereits die vierte A6-Generation. Sie trägt die interne Typenbezeichnung „4G", und kam 2011 als Limousine und Avant auf dem Markt. Der A6 C7 wird, wie zuvor schon der Audi A4, A5 und A7, nach dem Prinzip des Modularen Längsbaukastens (MLB) zusammengestellt. Darunter versteht man ein 2007 von Audi entwickeltes Plattformkonzept mit längs zur Fahrtrichtung verbautem Antriebsstrang. Dies führte zu einem Platztausch von Differenzial und Kupplung, wodurch die Vorderachse um rund 12 Zentimeter nach vorn rückte. Dies bescherte der vierten Generation des A6 einen längeren Radstand und eine ausgewogenere Gewichtsverteilung. Der A6 C7 besteht zum Teil aus Aluminium, andere Bauteile werden aus hoch- und höherfestem Stahl gefertigt. Das sorgt für eine Gewichtsreduktion um bis zu 70 Kilogramm.

Specifications

Manufacturer:	Audi
Country:	Germany
Model:	A6 3.0 TDI V6
Model:	Sedan
Produced:	since 2010
Length:	4933 mm
Width:	1874 mm
Height:	1455 mm
Wheelbase:	2912 mm
Empty weight:	1845 kg
Type of drive:	All-wheel drive
Max. speed:	250 km/h (regulated)
Fuel consumption:	5.1 l/100 km

*data for the 2014 Audi A6 3.0 TDI V6 2014

Today, numerous cars are built using a modular construction system, which helps to save development and production costs. This also applies to the Audi A6 C7, the fourth generation of the A6. Internally designated as "4G," it was introduced as sedan and as a station wagon called Avant. As the A4, the A5 and the A7, the A6 C7 is built based on the principle of the modular longitudinal construction kit (Modularer Längsbaukasten, MLB), a platform concept developed by Audi in 2007 with a power train arranged in the direction of travel. Due to this concept, the differential and the clutch had to swap places and the front axle had to move 12 cm forwards. This resulted in a longer wheelbase and a more balanced weight distribution of the fourth generation of the A6. The A6 C7 is made in parts of aluminum, while other parts are manufactured of high-strength steel. This causes a reduction in weight by up to 70 kg.

Der Motor

Motordaten

Bauart:	Diesel-90-Grad-V-Motor mit Turbolader
Zylinderzahl:	6
Hubraum:	2.967 cm³
Bohrung:	83 mm
Hub:	91,4 mm
Leistung:	200 kW (272 PS) 3.500–4.250/min

Der 3-Liter-90-Grad-V6 lässt sich als wahres Kraftpaket bezeichnen. Im weiten Drehzahlbereich von 3.500 bis 4.250 Umdrehungen pro Minute gibt er seine höchste Leistung von 200 kW (272 PS) ab. Imposant ist auch sein maximales Drehmoment von 580 Nm, das von 1.250 bis 3.250 Touren anliegt. Für die Gemischaufbereitung kommt ein Common-Rail-Einspritzsystem mit Piezo-Injektoren und einem Druck von maximal 2.000 bar zum Einsatz. Zu den weiteren Merkmalen zählen Achtloch-Düsen, eine VTG-Turboaufladung mit Ladeluftkühlung sowie Drall- und Tangential-Einlasskanäle. Jeder der sechs Brennräume verfügt über zwei Ein- und Auslassventile. Sie werden über je zwei oben liegende Nockenwellen und Rollenschlepphebel mit hydraulischem Ventilspielausgleich betätigt. Den 3-Liter-TDI-V6-Motor gibt es auch in einer leistungsreduzierten Variante mit 160 kW (218 PS) zwischen 4.000 und 4.750 Umdrehungen pro Minute. Das Drehmoment liegt bei 400 Nm zwischen 1.250 und 3.750 Touren.

The engine

Engine specifications

Type:	90° V Diesel engine with turbo charger
Number of cylinders:	6
Displacement:	2967 cm³
Bore:	83 mm
Stroke:	91.4 mm
Power:	200 kW (272 hp) 3500–4250 rpm

The 3-liter 90° V6 proves to be a real power pack. It delivers its maximum power of 200 kW (272 hp) in a wide speed range from 3500 to 4250 rpm. Another impressive feature is the maximum torque of 580 Nm at 1250 to 3250 rpm. Mixture is provided by a common rail injection system with piezo injectors and a pressure up to 2000 bar. Additional features consist of eight-hole jets, VTG turbo charging with charge air cooling as well as angular and tangential inlet canals. Each of the six combustion chambers has two intake and outlet valves controlled by two overhead camshafts each and roller cam followers with hydraulic valve-clearance compensation. There is also a reduced-power 3-liter TDI V6 engine with 160 kW (218 hp) from 4000 to 4750 rpm. The maximum torque of this version amounts to 400 Nm and is reached from 1250 to 3750 rpm.

Seit 2014 gibt es die neue Generation des 3.0 TDI V6 mit 160 kW (218 PS) sowie 200 kW (272 PS).
Since 2014, the new generation of the 3.0 TDI V6 engine with 160 kW (218 hp) or 200 kW (272 hp) is available.

Der Audi A6 C7 wurde nach dem Prinzip des Modularen Längsbaukastens (MLB) konzipiert – einem 2007 von Audi entwickelten Plattformkonzept mit längs zur Fahrtrichtung verbautem Antriebsstrang.

The Audi A6 C7 is built based on the principle of the modular longitudinal construction kit, a platform concept developed by Audi in 2007 with a power train arranged in the direction of travel.

Der Bentley Continental GT V8
vereint britische Automobilkultur
mit Sportwagendynamik.

The Bentley Continental GT V8
combines British automobile culture
and sports car dynamics.

Bentley Continental GT V8

Fahrzeugdaten

Hersteller:	Bentley
Land:	Großbritannien
Modell:	Continental GT V8
Bauzeit:	2012–2015
Länge:	4.806 mm
Breite:	1.944 mm
Höhe:	1.394 mm
Radstand:	2.746 mm
Leergewicht:	2.295 kg
Antriebsart:	Allrad
Höchstgeschwindigkeit:	303 km/h
Verbrauch:	10,6 Liter/100 km

Die Marke Bentley ist neben Rolls-Royce der Inbegriff britischer Noblesse. Doch der Continental GT V8 war weniger Brite, als es den Anschein hatte. So war bereits die erste bis 2011 gebaute Version der erste Bentley, der unter der Regie von Volkswagen entstand. Was jedoch stets britisch blieb, war die Handwerkskunst, mit welcher der Wagen mit viel Liebe zum Detail zusammengebaut wurde. Seit April 2012 gab es den bis dahin nur mit einem W12-Motor angebotenen Briten auch mit einem V8-Biturbo-Motor. Der V8-Motor war von sportlichem Charakter und stand dem W12 kaum nach. Den Sprint von 0 auf 100 km/h erledigte er sogar etwas schneller. Insgesamt überzeugten die Fahrleistungen des Bentley restlos, waren sie doch besser als bei manchem Sportwagen. Nahm man im Cockpit des Continental GT Platz, wähnte man sich dagegen eher in einem rollenden Luxus-Wohnzimmer und vergaß nahezu, dass in dem Wagen das Feuer eines reinrassigen Supersportlers loderte.

Specifications

Manufacturer:	Bentley
Country:	Great Britain
Model:	Continental GT V8
Produced:	2012–2015
Length:	4806 mm
Width:	1944 mm
Height:	1394 mm
Wheelbase:	2746 mm
Empty weight:	2295 kg
Type of drive:	All-wheel drive
Max. speed:	303 km/h
Fuel consumption:	10.6 l/100 km

Together with Rolls-Royce, Bentley is the epitome of British noblesse. However, the Continental GT V8 was less British than it seemed. The first version built until 2011 was the first Bentley created under the direction of Volkswagen. The craftsmanship, on the other hand, remained very British, so that the cars were built with much attention to detail. The car was originally only offered with a W12 engine, but starting with April 2012, a V8 twin turbo engine was also available. This engine was rather sporty and in most aspects did not rank behind the W12. It even needed a little less time to accelerate from 0 to 100 km/h. The driving characteristics of the Bentley were highly convincing and even better than that of some sports cars. When you took a seat in the cockpit of a Continental GT, you could imagine to sit in a mobile luxury lounge and nearly forget the fire of a pure sports car burning within.

Der Motor

Motordaten

Bauart:	Viertakt-90-Grad-V-Motor
Zylinderzahl:	8
Hubraum:	3.993 cm³
Bohrung:	84,5 mm
Hub:	89 mm
Leistung:	373 kW (507 PS)
	bei 6.000/min

Der V8-Biturbo-Motor des Bentley Continental GT V8 kam aus Ingolstadt. In seiner Basisversion bewegte er den 2,5 Tonnen schweren Wagen mit 373 kW (507 PS), die er bei 6.000 Touren bereitstellte. Sein maximales Drehmoment von 660 Nm gab das Aggregat bereits ab 1.700 Umdrehungen pro Minute ab. Jeder Zylinder war mit je zwei Ein- und Auslassventilen ausgestattet, die über zwei oben liegende Nockenwellen gesteuert wurden. Der Motor wirkte über ein Achtgang-Automatikgetriebe auf einen Allradantrieb, der 40 Prozent des Drehmoments auf die Vorder- und 60 Prozent auf die Hinterräder leitete. Binnen 4,8 Sekunden wurde aus dem Stillstand Tempo 100 erreicht. Nachdem der Bentley Continental GT V8 innerhalb des ersten Jahres genauso oft verkauft worden war wie die Continental-GT-Modelle mit mächtigem W12-Motor, bot man den bekannten V8-Motor auch in einer weiteren Leistungsstufe an. Im GT V8 S lieferte er 389 kW (528 PS) und bot ein um 20 Nm höheres Drehmoment.

The engine

Engine specifications

Type:	Four-stroke 90° V engine
Number of cylinders:	8
Displacement:	3993 cm³
Bore:	84.5 mm
Stroke:	89 mm
Power:	373 kW (507 hp)
	at 6000 rpm

The V8 twin turbo engine of the Bentley Continental GT V8 was made in Ingolstadt. In the base version, it moved the 2.5 ton vehicle with 373 kW (507 hp) at 6000 rpm. The maximum torque of 660 Nm was reached at 1700 rpm. Each cylinder was equipped with two intake and outlet valves, which were controlled by two overhead camshafts. The engine power was transferred via an 8-gear automatic transmission to an all-wheel drive, which distributed 40 % of the torque to the front wheels and 60 % to the rear wheels. Accelerating from 0 to 100 km/h took 4.8 seconds. In the first year, as many units of the Bentley Continental GT V8 were sold as of the Continental GT models with the powerful W12 engine. Thus the V8 was also offered with a different power rating. In the GT V8 S, it delivered 389 kW (528 hp) and provided 20 Nm more torque.

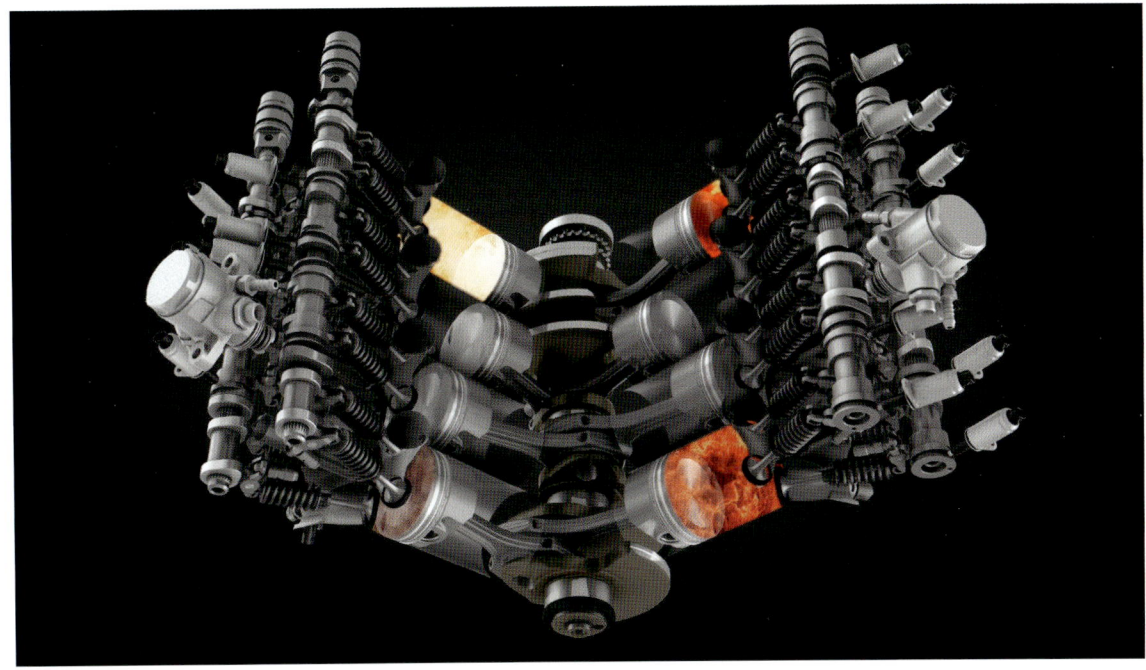

Auch mit dem 4-Liter-V8-
Biturbo-Motor wurde der
Bentley Continental GT V8
zum Verkaufsschlager.

The Bentley Continental GT V8
with 4-liter V8 twin turbo engine
also became a bestseller.

Der 4-Liter-V8-Motor übertrug
sein Drehmoment über ein
Achtgang-Automatikgetriebe.

The 4-liter V8 engine transferred
its torque via an automatic
transmission.

Bei Bentley gehen hohe sportliche Performance und gediegene britische
Handwerkskunst eine perfekte Symbiose ein. Das Ergebnis sind stets außergewöhnlich hochwertige Automobile.

Bentley perfectly combines sporty performance and solid British craftsmanship, which always results in high-quality vehicles.

Supersportwagen: Ab 2013 baute McLaren den P1 in einer limitierten Auflage von 375 Exemplaren.
Super sports car: From 2013 on, McLaren built the P1 in a limited production run of 375 units.

McLaren P1

Fahrzeugdaten

Hersteller:	McLaren
Land:	Großbritannien
Modell:	P1
Bauzeit:	2013–2015
Länge:	4.588 mm
Breite:	1.946 mm
Höhe:	1.188 mm
Radstand:	2.680 mm
Leergewicht:	1395 kg
Antriebsart:	Hinterrad
Höchstgeschwindigkeit:	350 km/h
Verbrauch:	8,3 Liter/100 km

Der P1 war ein Hybrid-Sportwagen, den die britische Marke McLaren zwischen 2013 und 2015 in einer limitierten Auflage von 375 Exemplaren fertigte. Als Verbrennungsmotor kam im McLaren ein 542 kW (737 PS) starker 3,8-Liter-V8-Motor mit zwei Turboladern zum Einsatz. Zusätzlich verfügte der P1 über einen Elektromotor, der 132 kW (179 PS) leistete. Insgesamt produzierte dieser extreme Renner damit eine Gesamtleistung von 674 kW (916 PS). McLaren bezeichnete den P1 als eigentlichen Nachfolger des Modells F1, das in den 1990er-Jahren neue Maßstäbe im Bereich der Straßensportwagen gesetzt hatte. Zusätzlich hatte der McLaren F1 im Jahr 1995 die 24 Stunden von Le Mans gewonnen. Der P1 übertraf alle Fahrleistungen seines legendären Vorgängers deutlich. Von 0 auf 100 km/h benötigte der Hybrid-McLaren nur 2,8 Sekunden – und Tempo 200 war bereits nach 6,8 Sekunden erreicht. Die Höchstgeschwindigkeit des Supersport-wagens war elektronisch auf 350 km/h begrenzt.

Specifications

Manufacturer:	McLaren
Country:	Great Britain
Model:	P1
Produced:	2013–2015
Length:	4588 mm
Width:	1946 mm
Height:	1188 mm
Wheelbase:	2680 mm
Empty weight:	1395 kg
Type of drive:	Rear-wheel drive
Max. speed:	350 km/h
Fuel consumption:	8.3 l/100 km

The P1 was a hybrid sports car built by the British McLaren brand between 2013 and 2015 with a limited production run of 375 units. It was driven by a 3.8-liter V8 combustion engine with twin turbo charger and 542 kW (737 hp) and an additional electromotor with 132 kW (179 hp). In total, the engine power of this extreme racing car thus amounted to 674 kW (916 hp). McLaren called the P1 the true successor of the model F1, which had set new standards for street sports cars in the 1990s. The McLaren F1 had also won the 24 Hours of Le Mans in 1995. The P1 significantly exceeded the driving characteristics of its legendary predecessors. It only needed 2.8 seconds to accelerate from 0 to 100 km/h and only 6.8 seconds to reach 200 km/h. The top speed of the super racing car was electronically limited to 350 km/h.

Der Motor

Motordaten

Bauart:	Viertakt-90-Grad-V-Motor mit Turbolader / Elektromotor
Zylinderzahl:	8
Hubraum:	3.799 cm³
Bohrung:	93 mm
Hub:	69,9 mm
Leistung	
Verbrennungsmotor:	542 kW (737 PS) bei 7.500/min
Leistung Elektromotor:	132 kW (179 PS)
Gesamtleistung:	674 kW (916 PS)

Das V8-Biturbo-Triebwerk des McLaren P1 war als Mittelmotor vor der Hinterachse eingebaut. Der 3,8-Liter leistete 542 kW (737 PS) bei 7500 Umdrehungen pro Minute. Das maximale Drehmoment des Verbrennungsmotors betrug 720 Nm, die er bei 4000/min lieferte. Insgesamt basierte die Motorenkonstruktion grundsätzlich auf dem Antriebsaggregat des McLaren 12C, war allerdings in verschiedenen Details von den Ingenieuren überarbeitet worden. So kamen ein veränderter Motorblock sowie neue Turbolader zum Einsatz. Der Elektromotor mit seinen 132 kW (179 PS) war in den Antriebsstrang integriert. Mit seiner Hilfe ließen sich die ohnehin schon beeindruckenden Fahrleistungen des McLaren P1 noch weiter verbessern. Insgesamt lieferte der Hybrid-Sportwagen eine Gesamtleistung von 674 kW (916 PS).

Die Zusatzleistung des Elektromotors konnte per Knopfdruck oder automatisch abgerufen werden. Die Batterie wurde vom Verbrennungsmotor oder einer Plug-in-Ladestation aufgeladen.

The engine

Engine specifications

Type:	Four-stroke 90° V engine with turbo charger/electromotor
Number of cylinders:	8
Displacement:	3799 cm³
Bore:	93 mm
Stroke:	69.9 mm
Combustion engine power:	542 kW (737 hp) at 7500 rpm
Electromotor power:	132 kW (179 hp)
Total engine power:	674 kW (916 hp)

The 3.8-liter V8 twin turbo engine of the McLaren P1 was mounted as a midship engine in front of the rear engine and delivered 542 kW (737 hp) at 7500 rpm. The maximum torque of the combustion engine amounted to 720 Nm and was reached at 4000 rpm. The motor design was based on the engine of the McLaren 12C, albeit with modifications in many details. For instance, a modified engine block and new turbo chargers were used. The 132 kW (179 hp) electromotor was integrated in the power train and helped to further increase the impressive driving performance of the McLaren P1. The total power of the hybrid sports car amounted to 674 kW (916 hp). The additional power of the electromotor could be summoned automatically or by the touch of a button. The battery was charged by the combustion engine or a plug-in charger.

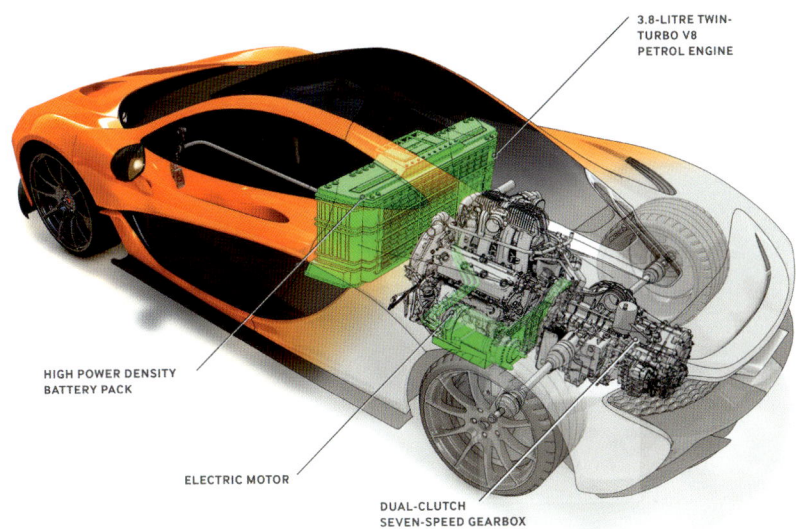

3.8-LITRE TWIN-
TURBO V8
PETROL ENGINE

HIGH POWER DENSITY
BATTERY PACK

ELECTRIC MOTOR

DUAL-CLUTCH
SEVEN-SPEED GEARBOX

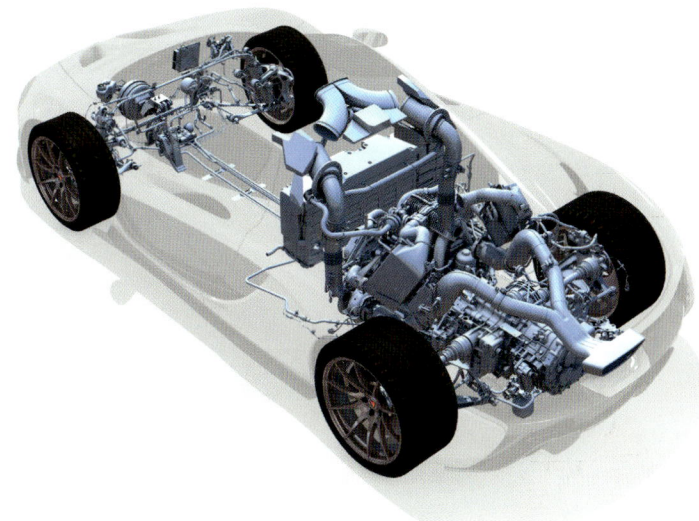

Der Hybridantrieb bescherte dem McLaren P1 eine Gesamtleistung von 674 kW (916 PS).
The hybrid drive gave the McLaren P1 a total power of 674 kW (916 hp).

Der Biturbo-V8 und der Elektromotor trieben gemeinsam die Hinterräder an.
The twin turbo V8 engine and the electromotor jointly drove the rear wheels.

Der 3,8-Liter-V8-Verbrennungsmotor
des McLaren P1 war mit zwei
Turboladern versehen.

The 3.8-liter V8 combustion engine
of the McLaren P1 was equipped
with two turbo chargers.

Die Kombination aus einem 3,8-Liter-V8-Biturbo-Triebwerk und einem Elektromotor bescherte dem P1 eine Leistung von 674 kW (916 PS). McLaren begrenzte die Höchstgeschwindigkeit des Hybrid-Renners auf 350 km/h.

The combination of a 3.8-liter V8 twin turbo engine and an electromotor gave the P1 a total power of 674 kW (916 hp). McLaren limited the top speed of the hybrid racing car to 350 km/h.

Links die Corvette Z06 (Straßenversion) von 2015
und rechts die Corvette C7.R aus dem Jahr 2014.

The 2015 Corvette Z06 (street version; left) and
the 2014 Corvette C7.R.

Chevrolet Corvette C7 Z06

Fahrzeugdaten

Hersteller:	Chevrolet
Land:	USA
Modell:	Corvette C7 Z06
Bauzeit:	seit 2013
Länge:	4.518 mm
Breite:	1.965 mm
Höhe:	1.235 mm
Radstand:	2.710 mm
Leergewicht:	1.598 kg
Antriebsart:	Hinterrad
Höchstgeschwindigkeit:	> 320 km/h
Verbrauch:	ca. 14 Liter/100 km

Wie kaum ein anderer Wagen schrieb die Chevrolet Corvette amerikanische Automobilgeschichte. Die Corvette wurde 1953 erstmals vorgestellt und hat seitdem bereits ihre siebte Auflage erfahren. Die letzte Neuauflage wurde 2013 vorgestellt. Über Jahrzehnte definierte sich das Aushängeschild der US-Cars eher durch spektakuläres Design und große Motoren. In den Fahrleistungen hinkte sie europäischen Sportwagen jedoch hinterher. Das änderte sich erst mit der Einführung der sechsten Generation 2005. Seitdem zählt die Corvette als ein zu europäischen Sportwagen konkurrenzfähiges Produkt.

Die Chevrolet Corvette C7 Z06, die siebte Generation, wurde gemeinsam mit dem C7.R-Rennwagen entwickelt. Ihr Chassis besteht aus Aluminium- und Magnesium-Komponenten. Bei der Karosserie setzte man auf Kunststoff und Kohlefaser. Die Corvette C7 Z06 unterscheidet sich vom Basismodell primär durch die kraftvollere Motorisierung. Während die C7 343 kW (466 PS) leistet, packt die Z06 rund 140 Kilowatt drauf.

Specifications

Manufacturer:	Chevrolet
Country:	USA
Model:	Corvette C7 Z06
Produced:	since 2013
Length:	4518 mm
Width:	1965 mm
Height:	1235 mm
Wheelbase:	2710 mm
Empty weight:	1598 kg
Type of drive:	Rear-wheel drive
Max. speed:	> 320 km/h
Fuel consumption:	approx. 14 l/100 km

The Chevrolet Corvette wrote American car history like nearly no other car. It was first introduced in 1953 and has already experienced a seventh generation, which was presented in 2013. For decades, the figurehead of US cars defined itself by a spectacular design and large engines. The driving performance, however, lagged behind that of European sports cars. This was only changed with the introduction of the sixth generation in 2005. Since then, the Corvette is deemed competitive with European sports cars.

The Chevrolet Corvette C7 Z06 was developed together with the C7.R racing car. The chassis is made of aluminum and magnesium components. For the auto body, the engineers relied on plastic and carbon fiber. The Corvette C7 Z06 is mainly distinguished from the base model by its more powerful engine. While the C7 provides 343 kW (466 hp), the Z06 delivers an additional 140 kW.

Der Motor

Motordaten

Bauart:	Viertakt-90-Grad-V-Motor mit Kompressor
Zylinderzahl:	8
Hubraum:	6.162 cm³
Bohrung:	103,25 mm
Hub:	92 mm
Leistung:	485 kW (659 PS) 6.400/min

Der 6,2-Liter-V8-Motor mit Kompressor ist mit 485 kW (659 PS) der leistungsstärkste Motor, der bis dahin in einer Serien-Corvette eingebaut wurde. Seine maximale Leistung stellt er bei einer Drehzahl von 6.400 Umdrehungen pro Minute bereit. Der Ladedruck beträgt 0,65 bar. Das maximale Drehmoment liegt bei 881 Nm bei 3.600 Touren. Seit 1955 ist die Corvette dem V8-Motor treu geblieben. Neu sind allerdings bei der seit 2013 verfügbaren siebten Generation die Benzin-Direkteinspritzung sowie eine variable Ventilsteuerung. Erstmals wird seitdem auch eine Zylinderabschaltung eingebaut. Sie koppelt bei geringer Motorlast vier Zylinder ab, sodass der Motor nur auf vier Zylindern läuft und dadurch Kraftstoff spart. Das aus Aluminium gefertigte Aggregat verfügt über eine Trockensumpfschmierung, eine zentral platzierte Nockenwelle, und jeder Zylinder ist mit einem Ein- und einem Auslassventil versehen. Das Drehmoment wirkt über ein Siebengang-Getriebe auf die beiden Hinterräder.

The engine

Engine specifications

Type:	Four-stroke 90° V engine with compressor
Number of cylinders:	8
Displacement:	6162 cm³
Bore:	103.25 mm
Stroke:	92 mm
Power:	485 kW (659 hp) at 6400 rpm

The 6.2-liter V8 engine with compressor provides 485 kW (659 hp) and is thus the most powerful engine that has been mounted in a production Corvette up today. It reaches its maximum power at 6400 rpm. The charging pressure amounts to 0.65 bar. The maximum torque of 881 Nm is reached at 3600 rpm. Since 1955, the Corvette remained loyal to the V8 engine. New features of the 7th generation introduced in 2013, however, are the direct gasoline injection and a variable valve control. Additionally, a cylinder cut-off was integrated for the first time. At low engine load, it disconnects four cylinders so that the engine only runs on four cylinders and thus saves fuel. The aluminum engine is equipped with dry sump lubrication, a central camshaft and one intake and outlet valve per cylinder. The torque is transferred to the rear wheels via a seven-gear transmission.

Der 6,2-Liter-V8-Smallblock-Motor der Corvette Z06 (2014) besitzt eine unten liegende Nockenwelle.
The 6.2-liter V8 small block engine of the Corvette Z06 (2014) has a bottom-mounted camshaft.

Die Chevrolet Corvette ist der Inbegriff des amerikanischen Sportwagens schlechthin. In puncto Tradition ist der Vergleich mit dem Porsche 911 angesichts der langen Produktionsdauer durchaus statthaft.

The Chevrolet Corvette is the epitome of American sports cars. Regarding tradition, it is comparable with the Porsche 911, given the long production period.

Der McLaren 12C wurde von 2011 bis 2014 gebaut – im Bild die Spider-Variante.
The McLaren 12C was built from 2011 to 2014. The figure shows the Spider version.

McLaren 12C

Fahrzeugdaten

Hersteller:	McLaren
Land:	Großbritannien
Modell:	12C
Bauzeit:	2011–2014
Länge:	4.509 mm
Breite:	1.908 mm
Höhe:	1.199 mm
Radstand:	2.670 mm
Leergewicht:	1.336 kg
Antriebsart:	Hinterrad
Höchstgeschwindigkeit:	333 km/h
Verbrauch:	11,7 Liter/100 km

*Werte für 2013 McLaren 12C

Der 12C war das erste Serienfahrzeug, das McLaren seit dem Produktionsende des legendären Modells F1 unter eigenem Namen auf den Markt brachte. Der Sportwagen wurde 2011 präsentiert, zunächst noch unter dem Namen „McLaren MP4-12C". Ab 2013 firmierte das Modell unter der Bezeichnung „12C". Der Wagen war als Coupé und als Spider erhältlich.

Die Historie der Marke begann 1963, als der Rennfahrer Bruce McLaren das Unternehmen in England gründete. Ab 1966 trat McLaren in der Formel 1 an. Schon zwei Jahre später feierte der Rennstall seinen ersten Grand-Prix-Sieg – und 1974 den ersten WM-Titel. Den ersten Straßensportwagen stellte McLaren 1993 auf die Räder.

Der Mittelmotor-Sportwagen 12C verfügte über ein Kohlefaser-Monocoque, ein Siebengang-Doppelkupplungsgetriebe und einen 3,8-Liter-V8-Biturbo-Motor. Das Triebwerk leistete 460 kW (625 PS) bei 7500/min. Für den Sprint von 0 auf 100 km/h benötigte der Supersportler 3,3 Sekunden. 2014 endete die Herstellung des 12C.

Specifications

Manufacturer:	McLaren
Country:	Great Britain
Model:	12C
Produced:	2011–2014
Length:	4509 mm
Width:	1908 mm
Height:	1199 mm
Wheelbase:	2670 mm
Empty weight:	1336 kg
Type of drive:	Rear-wheel drive
Max. speed:	333 km/h
Fuel consumption:	11.7 l/100 km

* data for the 2013 McLaren 12C

The 12C was the first production car that McLaren launched under its own brand after discontinuation of the legendary model F1. The sports car was introduced in 2011, first under the name of "McLaren MP4-12C." Beginning with 2013, the model took on the designation "12C." It was available as a coupé and as a Spider version.

The brand history began in 1963, when the race driver Bruce McLaren founded the company in England. In 1966, McLaren participated in the Formula 1 for the first time. Only two years later, the racing team could celebrate its first Grand Prix win. In 1974, the first world championship title followed. McLaren built its first street sports car in 1993.

The midship engine sports car 12C had a carbon fiber monocoque, a seven-gear dual-clutch transmission and a 3.8-liter V8 twin turbo engine with 460 kW (625 hp) at 7500 rpm. Accelerating from 0 to 100 km/h took the super sports car 3.3 seconds. Production of the 12 C ended in 2014.

Der Motor

Motordaten

Bauart:	Viertakt-90-Grad-V-Motor mit Turbolader
Zylinderzahl:	8
Hubraum:	3.799 cm³
Bohrung:	93 mm
Hub:	69,9 mm
Leistung:	460 kW (625 PS) bei 7.500/min

Der 3,8 Liter große Biturbo-V8-Motor mit der Typenbezeichnung „M838T" war eine Konstruktion, die in Zusammenarbeit zwischen McLaren und der britischen Firma Ricardo entstand. Zwischen dem Beginn des Projekts und dem Start der Produktion lagen lediglich 18 Monate.

Der Zylinderwinkel des Aggregats betrug 90 Grad. Der Vierventiler verfügte über vier oben liegende Nockenwellen. Zudem besaß er eine Trockensumpfschmierung. Der McLaren-Motor wog 200 Kilogramm und brachte es anfangs auf eine Leistung von 440 kW (600 PS). Ab Herbst 2012 leistete das Triebwerk 460 kW (625 PS) bei 7.500 Umdrehungen pro Minute. Die Höchstdrehzahl betrug 8.500/min. Im Bereich zwischen 3.000 und 7.000 Umdrehungen lieferte der M838T sein maximales Drehmoment von 600 Nm. Seine Kraft übertrug der Mittelmotor über ein Siebengang-Doppelkupplungsgetriebe auf die Hinterräder.

The engine

Engine specifications

Type:	Four-stroke 90° V engine with turbo charger
Number of cylinders:	8
Displacement:	3799 cm³
Bore:	93 mm
Stroke:	69.9 mm
Power:	460 kW (625 hp) at 7500 rpm

The 3.8-liter twin turbo V8 engine with the type designation "M838T" was developed in cooperation by McLaren and the British company Ricardo. It took only 18 months from the beginning of the project to the start of production.

The banking angle amounted to 90°. The four-valve engine had four overhead camshafts and used dry sump lubrication. The McLaren engine weighed 200 kg. At first, it provided 440 kW (600 hp). From autumn 2012 on, however, the engine delivered 460 kW (625 hp) at 7500 rpm, with a maximum engine speed of 8500 rpm and a maximum torque of 600 Nm in the range from 3000 to 7000 rpm. The power of the midship engine was transferred to the rear wheels via a seven-gear double-clutch transmission.

Das Triebwerk mit der internen Bezeichnung „M838T" kam im McLaren 12C als Mittelmotor zum Einsatz.

The engine with the internal designation "M838T" was mounted as a midship engine in the McLaren 12C.

Der 3,8-Liter-V8-Motor war eine McLaren-Eigenentwicklung, die erstmals im Modell 12C eingebaut wurde.

The 3.8-liter V8 engine is an in-house development by McLaren and was first used in the model 12C.

Der McLaren 12C besaß 2011 einen V8-Biturbo-Motor und ein Kohlefaser-Monocoque. Als erstes Team hatte McLaren schon im Jahr 1981 das Chassis eines Formel-1-Wagens aus Kohlefaser-Material hergestellt.

In 2011, the McLaren 12C had a V8 twin turbo engine and a carbon-fiber monocoque. McLaren had already been the first team to build the chassis of a Formula 1 car from carbon fiber in 1981.

Motoreinbau bei McLaren – die „Hochzeit"

Das McLaren Production Center, kurz MPC, ist am Stadtrand von London angesiedelt. In der rund 34.500 Quadratmeter großen Anlage werden High-Performance-Straßenautos zusammengebaut. McLaren hat sich auf die Entwicklung hochwertiger Sportwagen und ihrer Komponenten spezialisiert. Die Fertigung der einzelnen Bauteile überlässt die Sportwagenschmiede ausgesuchten Spezialfirmen, die ihr Handwerk bestens verstehen. In den Fabrikhallen von McLaren werden die zugelieferten Einzelteile zu edlen Sportwagen zusammengefügt. Dort dominiert Handarbeit. Sie ist der Schlüssel zu vollendeter Qualität. Der Höhepunkt bei der Herstellung eines jeden Wagens ist die Vermählung des Fahrgestells mit dem Antriebsaggregat. Dieser Akt wird von qualifizierten Technikern unter quasi Reinraum-Bedingungen in Handarbeit vollzogen. Dabei wird nichts dem Zufall überlassen. Gleichzeitig unterscheidet sich diese Art der Automobilfertigung erheblich von der automatisierten Massenproduktion üblicher PKW.

Engine mounting at McLaren – the "marriage"

The McLaren Production Center (MPC) is located in the outlying areas of London. The site of approx. 34,500 m^2 is used to assemble high-performance street cars. McLaren specializes in the development of premium sports cars and their components. Manufacturing is delegated to selected specialist companies who know their craft. The supplied components are then assembled into noble sports cars at the McLaren factories. Here, manual work prevails. It is the key to perfect quality. The climax of the car assembly is the marriage of the undercarriage with the engine. This act is done by qualified technicians nearly under clean room conditions. Nothing is left to chance. This type of car manufacturing is significantly different from the automated mass production of usual passenger cars.

Einbau eines 3,8-Liter-V8-Motors vom Typ M838T in einen McLaren MP-4 12C Spider.

Mounting a 3.8-liter V8 M838T engine in a McLaren MP-4 12C Spider.

Motoreinbau bei McLaren – die „Hochzeit"

Der Porsche 911 Carrera S und das 911 Carrera Cabriolet aus dem Jahr 2015.
The Porsche 911 Carrera S and the 911 Carrera cabriolet of the year 2015.

Porsche 911 Carrera S 2015

Fahrzeugdaten

Hersteller:	Porsche
Land:	Deutschland
Modell:	911 Carrera S
Baujahr:	seit 2011
Länge:	4.499 mm
Breite:	1.808 mm
Höhe:	1.296 mm
Radstand:	2.450 mm
Leergewicht:	1.515 kg
Antriebsart:	Hinterrad
Höchstgeschwindigkeit:	308 km/h
Verbrauch:	8,7 Liter /100 km

*Werte für 911 Carrera S (991.2) 2015

Die siebte Auflage des Porsche 911 vom Typ 991 wurde Ende 2011 eingeführt. Vier Jahre später folgte mit dem 991.2 eine überarbeitete Version. Erstmals in der Geschichte des Porsche 911 wird die Grundversionen nicht mehr mit einem Saug-, sondern mit einem Turbomotor ausgeliefert. Fand sich im 911 Carrera S noch ein 3,8-Liter-Motor mit 294 kW (400 PS), so leistet der Turbomotor des 2015er-Modells mit nur noch 3 Litern Hubraum 309 kW (420 PS). Weiter konnte der Benzinverbrauch um etwa 0,8 Liter reduziert werden. Es ist beeindruckend, wie sich der Porsche 911 von 1963 bis 2015 weiterentwickelt hat. Betrachten wir die Sechszylinder-Boxermotoren von einst und heute: Sie sind in puncto Hubraum zwar nur um 50 Prozent gewachsen, geben aber mehr das Dreifache an Leistung ab. Auch in Sachen Höchstgeschwindigkeit hat sich einiges getan. Lag sie vor mehr als einem halben Jahrhundert bei etwas mehr als 200 km/h, so sind es heute über 300 Stundenkilometer.

Specifications

Manufacturer:	Porsche
Country:	Germany
Model:	911 Carrera S
Produced:	since 2011
Length:	4499 mm
Width:	1808 mm
Height:	1296 mm
Wheelbase:	2450 mm
Empty weight:	1515 kg
Type of drive:	Rear-wheel drive
Max. speed:	308 km/h
Fuel consumption:	8.7 Liter /100 km

* data for the 911 Carrera S (991.2) introduced in 2015

Type 991, the seventh edition of the Porsche 911, was introduced in the end of 2011. Four years later, the updated version 991.2 followed. For the first time in the history of the Porsche 911, the basic version is no longer shipped with a suction engine but with a turbo engine. While the 911 Carrera S still had a 3.8-liter engine with 294 kW (400 hp), the turbo engine of the 2015 model gains 309 kW (420 hp) from a displacement of a mere 3 liters. Additionally, the fuel consumption is reduced by 0.8 liters. It is impressive to see the development of the Porsche 911 from 1963 to 2015. The displacement of the 6-cylinder flat engines has only increased by half, but they now provide more than three times of the original power. The top speed has also improved significantly. More than half a century ago, it amounted to slightly more than 200 km/h, but today it is more than 300 km/h.

Der Motor

Motordaten

Bauart:	Viertakt-Boxermotor mit Turbolader
Zylinderzahl:	6
Hubraum:	2.981 cm³
Bohrung:	91 mm
Hub:	76,4 mm
Leistung:	309 kW (420 PS) bei 6.500/min

Wie beim ersten 911 aus dem Jahr 1963 pulsiert auch in der siebten Generation ein Sechszylinder-Boxermotor. Allerdings hat nur noch wenig mit dem Original von damals zu tun. Seit dem 991.2 von 2015 verfügen mit Ausnahme der GT3-Modelle alle Triebwerke über Turboaufladung. Im 911 Carrera S und 4S ist es ein 3-Liter-Boxer, der bei 6.500 Umdrehungen pro Minute 309 kW (420 PS) entfaltet. Der Kraftprotz stellt sein höchstes Drehmoment von 500 Nm von 1.700 bis 5.000 Touren bereit. Der Motor ist in Vierventil-Technik aufgebaut. Die Ventile werden pro Zylinderbank über zwei Nockenwellen gesteuert. Hinzu kommen eine Nockenwellenverstellung und eine Ventilhubschaltung sowie ein automatischer Ventilspielausgleich. Die Motoraufladung erfolgt über je zwei Abgasturbolader und Ladeluftkühler. Den größten Sechszylinder-Boxer findet man im 2015er-911-Turbo-S: Aus 3,8 Liter Hubraum schöpft er eine Leistung von gewaltigen 427 kW (580 PS).

The engine

Engine specifications

Type:	Four-stroke flat engine with turbo charger
Number of cylinders:	6
Displacement:	2981 cm³
Bore:	91 mm
Stroke:	76.4 mm
Power:	309 kW (420 hp) at 6500 rpm

Like the first 911 of the year 1963, the 7th generation is still powered by a 6-cylinder flat engine. Of course, it doesn't bear many similarities to the original version. Beginning with the 991.2 introduced in 2015 all engines with the exception of the GT3 models are turbo charged. In the 911 Carrera S and the 4S, a 3-liter flat engine provides 309 kW (420 hp) at 6500 rpm and a maximum torque of 500 Nm between 1700 and 5000 rpm. The engine uses a four-valve technology with the valves being controlled by two camshafts per cylinder bank. Additionally, the engine is equipped with cam shaft adjustment, valve stroke control and automatic valve-clearance compensation. Charging is done by two exhaust turbo chargers with charge air cooling. The largest 6-cylinder flat engine is used in the 911 Turbo S model of the year 2015, which gains a whopping 427 kW (580 hp) from a displacement of 3.8 liters.

Die Phantomgrafik des Porsche 911 Carrera S lässt die Heckmotoranordnung gut erkennen.

The phantom diagram of the Porsche 911 Carrera S shows the rear mounting of the engine.

Der Sechszylinder-Biturbo-Boxer des 911 Carrera S leistet mit 3 Litern Hubraum 309 kW (420 PS).

The 6-cylinder twin turbo flat engine of the 911 Carrera S has a displacement of 3 liters and provided 309 kW (420 hp).

Mit der zweiten Generation des 911 vom Typ 991 hielten auch bei Porsche das Downsizing und damit kleinvolumige, aber dennoch leistungsstarke und effiziente Turbomotoren Einzug.

With the second generation of the 911, type 991, downsizing and thus small and yet powerful and efficient turbo engines found their way to Porsche.

Mit der zweiten Generation des R8 (R8 V10/V10 plus) setzt Audi die Erfolgsstory des Supersportlers fort.
With the second generation of the R8 (R8 V10/V10 plus), Audi continued the success story of the super sports car.

Audi R8 V10 plus

Fahrzeugdaten

Hersteller:	Audi
Land:	Deutschland
Modell:	R8 V10 plus
Bauzeit:	seit 2015
Länge:	4.426 mm
Breite:	1.940 mm
Höhe:	1.240 mm
Radstand:	2.650 mm
Leergewicht:	1.555 kg
Antriebsart:	Allrad
Höchstgeschwindigkeit:	330 km/h
Verbrauch:	12,3 Liter/100 km

Mit dem R8 stellte Audi 2006 einen Mittelmotor-Sportwagen vor. Mit der Typenbezeichnung „R8" zielte Audi auf den gleichnamigen zwischen 1999 und 2005 gebauten Rennprototypen ab, der beim 24-Stundenrennen von Le Mans fünf Siege erringen konnte. Die Straßenvariante des R8 hat außer dem Namen jedoch kaum Gemeinsamkeiten mit den Rennboliden. Anfangs war der Wagen ausschließlich als Coupé erhältlich, seit 2010 ist er auch als Spyder auf dem Markt. Bis Ende 2014 wurden vom R8 etwas mehr als 26.000 Einheiten produziert. Der R8 setzt in weiten Teilen auf Leichtbauwerkstoffe wie Aluminium und Magnesium. Anfangs war der R8 nur mit V8-Motor erhältlich. 2015 erschien die zweite Generation des Audi R8. Sie trägt die interne Typenbezeichnung 4S. Der neue R8 nutzt dieselbe Plattform wie der Lamborghini Huracàn. Im Vergleich zur ersten Generation wirkt die Karosserie kantiger und dynamischer. Zudem ist sie leichter und stabiler. Der neue R8 ist der bis heute leistungsstärkste Serien-Audi.

Specifications

Manufacturer:	Audi
Country:	Germany
Model:	R8 V10 plus
Produced:	since 2015
Length:	4426 mm
Width:	1940 mm
Height:	1240 mm
Wheelbase:	2650 mm
Empty weight:	1555 kg
Type of drive:	All-wheel drive
Max. speed:	330 km/h
Fuel consumption:	12.3 l/100 km

In 2006, Audi introduced the new midship engine sports car R8. The type designation "R8" harkes back to the racing car prototype of the same name, which was built between 1999 and 2005 and gained five wins at the 24 Hours of Le Mans. Apart from the name, however, the street car version of the R8 doesn´t have many similarities to the racing car. At first, the car was only available as a coupé, but from 2010 on also as a roadster called "Spyder." Until the end of 2014, a little more than 26,000 R8 were produced. Most parts of the R8 are made from light materials like aluminum or magnesium. At first, the R8 was only available with a V8 engine. The second generation of the Audi R8 with the internal designation 4S was introduced in 2015. It uses the same platform as the Lamborghini Huracàn. Compared to the first generation, the auto body looks more angular and dynamic. It also weighs less and is more robust. The R8 is the most powerful production car that Audi had ever built up to that time.

Der Motor

Motordaten

Bauart:	Viertakt-90-Grad-V-Motor
Zylinderzahl:	10
Hubraum:	5.204 cm³
Bohrung:	84,5 mm
Hub:	92,8 mm
Leistung:	449 kW (610 PS)
	bei 8.250/min

Anfangs wurde der R8 mit einem 4,2 Liter großen V8-Motor mit Direkteinspritzung und einer Leistung von 309 kW (420 PS) angeboten. Er war zuvor schon im Audi RS4 zum Einsatz gekommen. 2009 wurde auch ein V10-FSI-Motor verfügbar, der sich bereits im Audi S6 bewährt hatte. Dieser 5,2 Liter große V10-Motor entwickelte eine Leistung von 386 kW (525 PS) und wurde für die zweite Generation des R8 weiterentwickelt. Der 90-Grad-V10 ist mit Trockensumpfschmierung ausgestattet, pro Zylinder sind je zwei Ein- und Auslassventile eingebaut, die über zwei oben liegende Nockenwellen betätigt werden. Bei 8.250 Umdrehungen pro Minute entwickelt er 449 kW (610 PS), das höchste Drehmoment von 560 Nm liegt bei 6.500 Touren an. Die Gemischaufbereitung erfolgt über ein kombiniertes Einspritzsystem aus Direkt- und Saugrohreinspritzung. Der R8 V10 plus ist mit einem automatisierten Siebenganggetriebe ausgestattet und beschleunigt in 3,2 Sekunden von 0 auf 100 km/h.

The engine

Engine specifications

Type:	Four-stroke 90° V engine
Number of cylinders:	10
Displacement:	5204 cm³
Bore:	84.5 mm
Stroke:	92.8 mm
Power:	449 kW (610 hp)
	at 8250 rpm

The R8 was first available with a 4.2-liter V8 engine with direct injection and 309 kW (420 hp), which had already been used in the Audi RS4. From 2009, the V10 FSI engine previously used in the Audi S6 was also available. This 5.2-liter V10 engine delivered 386 kW (525 hp) and was further developed for the second generation of the R8. The engine has a banking angle of 90°, uses dry sump lubrication and is equipped with two intake and outlet valves each per cylinder, which are controlled by two overhead camshafts. The engine provides 449 kW (610 hp) at 8250 rpm and reaches its maximum torque of 560 Nm at 6500 rpm. The mixture is provided by a combined direct and suction pipe injection system. The R8 V10 plus is equipped with an automatic 7-gear transmission and accelerates from 0 to 100 km/h in 3.2 seconds.

In der Durchsicht ist die Mittelmotoranordnung des Audi R8 V10 plus sehr schön zu erkennen.

This diagram shows the midship engine installation of the Audi R8 V10 plus.

Der 5,2-Liter-V10-Motor des Audi R8 V10 plus leistet gewaltige 449 kW (610 PS).

The 5.2-liter V10 engine of the Audi R8 V10 plus provides a whopping 449 kW (610 hp).

Das Antriebskonzept des Audi R8 V10 plus setzt auf die bewährte Allradtechnik „quattro".

The drive concept of the Audi R8 V10 plus relies on the proven "quattro" all-wheel technology.

Im Sinne einer bestmöglichen und vor allem auch ökonomischen Konzernentwicklung teilt sich der Audi R8 V10 plus einen Großteil seiner Technik mit dem noch exklusiveren Lamborghini Huracàn.

For the sake of the best possible and most economic development of the group, the Audi R8 V10 plus shares a huge part of its technology with the even more exclusive Lamborghini Huracàn.

Der Bentley Bentayga ist ein 301 km/h schneller Geländewagen der absoluten Luxusklasse.
The Bentley Bentayga is a luxury-class off-road car with a top speed of 301 km/h.

Bentley Bentayga

Fahrzeugdaten

Hersteller:	Bentley
Land:	England
Modell:	Bentayga
Bauzeit:	seit 2015
Länge:	5.140 mm
Breite:	1.998 mm
Höhe:	1.742 mm
Radstand:	2.995 mm
Leergewicht:	2.440 kg
Antriebsart:	Allrad
Höchstgeschwindigkeit:	301 km/h
Verbrauch:	19 Liter/100 km

Mit dem 2015 präsentierten Bentayga stellt Bentley das schnellste SUV der Welt: Die Höchstgeschwindigkeit des 2,4-Tonnen-Kolosses beträgt satt 301 km/h. Entsprechend wartet der Bentayga nicht nur mit Offroad-Qualitäten auf, sondern vermag auch eingefleischte Sportwagen-Fahrer das Fürchten zu lehren. Der Sprint von 0 auf 100 km/h in 4,1 Sekunden macht dies deutlich.

Konkurrenten, wie der Porsche Cayenne Turbo S oder der Range Rover Sport SVR, können dem Bentayga im Hinblick auf die Performance nicht das Wasser reichen. Genau genommen schlägt im Bentayga auch ein deutsches Herz. Schließlich basiert er auf der MLB-EVO-Plattform von Volkswagen, auf die auch schon die zweite Generation des Audi Q7 zurückgegriffen hatte. Dass der Bentayga kein reinrassiger Brite ist, offenbart sich schließlich auch darin, dass seine Aluminium-Karosserie in der slowakischen Hauptstadt Bratislava gebaut wird. Erst die Endmontage erfolgt im englischen Crewe.

Specifications

Manufacturer:	Bentley
Country:	England
Model:	Bentayga
Produced:	since 2015
Length:	5140 mm
Width:	1998 mm
Height:	1742 mm
Wheelbase:	2995 mm
Empty weight:	2440 kg
Type of drive:	All-wheel drive
Max. speed:	301 km/h
Fuel consumption:	19 l/100 km

With the Bentayga introduced in 2015, Bentley presented the world's fastest SUV. The top speed of this 2.4 t colossus amounts to a whopping 301 km/h. Accordingly, the Bentayga doesn´t only provide off-road characteristics, but can even scare the living daylights out of die-hard sports car fans, as the acceleration from 0 to 100 km/h in just 4.1 seconds shows. Competitors like the Porsche Cayenne Turbo S or Range Rover Sport SVR can´t hold a candle to the Bentayga when it came to performance. Strictly speaking, the Bentayga is German at heart. After all, it is based on the MLB-EVO platform by Volkswagen, which has already been used for the second generation of the Audi Q7. The fact that the aluminum auto body is built in the Slovakian capital Bratislava also proves that the Bentayga is not a pure-bred British citizen. The finally assembly is then done in Crewe, England.

Der Motor

Motordaten

Bauart:	Viertakt-W-Motor
Zylinderzahl:	12
Hubraum:	5.952 cm³
Bohrung:	84 mm
Hub:	89,5 mm
Leistung:	447 kW (608 PS)
	bei 5.000–6.000/min

Im Bentley Bentayga arbeitet ein 12-Zylinder-W-Motor mit 6 Litern Hubraum. Unter einem W-Motor versteht man grundsätzlich einen Hubkolbenmotor mit drei Zylinderbänken, die jeweils im selben Bankwinkel zueinander stehen. Beim W12 des Bentley verhält es sich jedoch anders, da er aus zwei VR6-Konzepten zusammengestellt wurde. In jeder dieser VR6-Einheiten stehen die Zylinder im Bankwinkel von 15 Grad zueinander. Die beiden VR6-Einheiten bilden schließlich einen Winkel von 72 Grad, was zu einem überaus kompakten Zwölfzylindermotor führt. In jedem seiner zwölf Zylindereinheiten arbeiten je zwei Ein- und Auslassventile, die über zwei oben liegende Nockenwellen pro Zylinderbank gesteuert werden. Die maximale Leistung von 447 kW (608 PS) gibt der Motor bei 5.000 bis 6.000 Umdrehungen pro Minute ab. Das maximale Drehmoment von 900 Nm steht von 1.350 bis 4.500 Touren zur Verfügung. Die Kraft des W12-Motors gelangt über eine Achtgang-Automatik auf alle vier Räder.

The engine

Engine specifications

Type:	Four-stroke W engine
Number of cylinders:	12
Displacement:	5952 cm³
Bore:	84 mm
Stroke:	89.5 mm
Power:	447 kW (608 hp)
	at 5000–6000/min

The Bentley Bentayga is driven by a 12-cylinder W engine with a displacement of 6 liters. In general, a W engine is a piston engine with three cylinder banks, all arranged at the same banking angle. However, the Bentley W12 is different as it is a combination of two VR6 units. In each of the units, the cylinders are arranged in a banking angle of 15°, while the two units form an angle of 72°. This results in a very compact 12-cylinder engine. There are two inlet and two outlet valves per cylinder, controlled by two overhead camshafts per cylinder bank. The maximum power of 447 kW (608 hp) is reached at 5000 to 6000 rpm, while the maximum torque of 900 Nm is available from 1350 to 4500 rpm. An 8-gear automatic transmission transfers the power of the W12 engine to all four wheels.

Im Bentley Bentayga arbeitet ein hochinteressanter 6-Liter-W12-Motor mit 447 kW (608 PS).

The heart of the Bentley Bentayga is a highly interesting 6-liter W12 engine with 447 kW (608 hp).

Der Antriebsstrang des Bentley Bentayga mit dem mächtigen W12-Motor als Kraftquell.

The power train of the Bentley Bentayga with the powerful W12 as its power source.

Mit dem Bentayga brachte Bentley 2015 das mit 301 km/h schnellste SUV auf den Markt. Der mächtige W12-Motor ist eine technische Besonderheit und mit 608 PS zudem kräftig geraten.

In 2015, Bentley introduced the fastest SUV to the market – the Bentayga with a top speed of 301 km/h. The powerful W12 is a technological peculiarity and with its 608 hp a powerful one at that.

Dem Land Rover Defender 110 (hier in der Ausführung seit 2012) war kein Gelände zu schwierig.
No terrain was too rough for the Land Rover Defender 110 (here, the version built since 2012).

Land Rover Defender 110 SW

Fahrzeugdaten

Hersteller:	Jaguar Land Rover
Land:	Großbritannien
Modell:	Land Rover Defender 110 SW
Bauzeit:	2012–2016
Länge:	4.785 mm
Breite:	1.790 mm
Höhe:	ca. 2.100 mm
Radstand:	2.794 mm
Leergewicht:	2.064 kg (5-Sitzer)
Antriebsart:	Allrad
Höchstgeschwindigkeit:	144 km/h
Verbrauch:	11,1 Liter/100 km

Der Land Rover ist der Inbegriff eines Geländewagens. Er wurde seit 1948 gebaut und bestach über Jahrzehnte hinweg mit seiner nüchtern praktischen Kastenform. Bei diesem Geländewagen, der bis 1990 einfach als Land Rover ohne weitere Typenbezeichnung angeboten wurde, ging es allein um optimale Geländegängigkeit. Der Land Rover war das Auto, mit dem sich noch so unwegsame Ziele erreichen ließen. Er kam überall dort zum Einsatz, wo es schwierigstes Gelände zu bewältigen galt. Auf ihn vertraute man bei Expeditionen in die entlegensten Winkel unserer Erde ebenso wie beim Militär. Seit 1990 wurde der Geländewagen als Defender bezeichnet. Ihn gab es in drei Grundvarianten, die entsprechend ihren Radständen in Zoll als Defender „90", „110" und „130" bekannt wurden. Bis zur Einführung des Namens „Defender" wurde der 90-Zöller als „Ninety" und der 110er als „One-Ten" bezeichnet. Am 29. Januar 2016 lief nach mehr als zwei Millionen Einheiten der letzte Defender vom Band.

Specifications

Manufacturer:	Jaguar Land Rover
Country:	Great Britain
Model:	Land Rover Defender 110 SW
Produced:	2012–2016
Length:	4785 mm
Width:	1790 mm
Height:	approx. 2100 mm
Wheelbase:	2794 mm
Empty weight:	2064 kg (5-seater)
Type of drive:	All-wheel drive
Max. speed:	144 km/h
Fuel consumption:	11.1 l/100 km

The Land Rover is the epitome of an off-road vehicle. It was built since 1948, and its practical box shape stood out positively for decades. This off-roader, which was offered without any further type designations until 1990, was all about suitability for rough terrain. The Land Rover was the car that could even reach the most wayless destinations. It was used whenever difficult terrain had to be conquered. Scientific expeditions to the farthest corners of the Earth as well as the military relied on it. Beginning with 1990, the off-road car was called Defender. It was available in three basis versions, designated according to their wheelbase in inch as Defender "90," "110" and "130." Until the introduction of the name Defender, the 90" and the 110" version had been called "Ninety" or "One-Ten," respectively. On January 29th 2016, the last of more than 2 million Defenders left the assembly line.

Der Motor

Motordaten

Bauart:	Viertakt-Diesel-Reihenmotor
Zylinderzahl:	4
Hubraum:	2.198 cm³
Bohrung:	86 mm
Hub:	94,6 mm
Leistung:	90 kW (122 PS)
	bei 3.500/min

Der Standardantrieb des Defender war seit jeher der Dieselmotor. Seit 2012 wurde er in Deutschland ausschließlich mit einem 2,2-Liter-Aggregat angeboten. Rund 90 Prozent des maximalen Drehmoments von 360 Nm wurden im relevanten Drehzahlbereich von 2.200 bis über 4.350 Umdrehungen pro Minute abgegeben. Damit schaffte die Maschine ideale Voraussetzungen, um auch unter schwierigsten Bedingungen zu überzeugen. Seine maximale Leistung von 90 kW (122 PS) gab das Aggregat bei 3.500 Umdrehungen pro Minute ab. Das Verdichtungsverhältnis lag bei 15,6:1. Der Zylinderblock wurde aus Gusseisen gefertigt. Als Zylinderkopfmaterial kam Aluminium zum Einsatz.

Der Motor wirkte über ein Sechsgang-Schaltgetriebe mit Geländeuntersetzung auf alle vier Räder. Damit ließen sich Steigungen oder Gefälle bis zu einem Winkel von 45 Grad überwinden. Weiterhin war das Mittendifferenzial sperrbar, um Radschlupf weitgehend zu vermeiden und die Traktion weiter zu verbessern.

The engine

Engine specifications

Type:	Four-stroke in-line Diesel engine
Number of cylinders:	4
Displacement:	2198 cm³
Bore:	86 mm
Stroke:	94.6 mm
Power:	90 kW (122 hp)
	at 3500 rpm

The standard drive of the Defender had always been a Diesel engine. Beginning with 2012, it was only available with a 2.2-liter engine in Germany. Approx. 90 % of the maximum torque of 360 Nm were delivered in the relevant engine speed range from 2200 to more than 4350 rpm. This laid the ideal foundation for the convincing performance even under roughest conditions. The maximum power of 90 kW (122 hp) was reached at 3500 rpm. The compression ratio amounted to 15.6:1. The cylinder block was made from cast iron. As material for the cylinder heads, aluminum was used.

The engine power was transferred to all four wheels via a six-gear transmission with off-road reduction gear. This made it possible to overcome rising and falling slopes up to 45°. Additionally, the middle differential could be locked to prevent wheel slip and to increase traction.

Von 2012 an wurde der Defender in Deutschland ausschließlich mit
2,2-Liter-Dieselmotor ausgeliefert.
From 2012 on, the Defender was shipped to Germany only with a
2.2-liter Diesel engine.

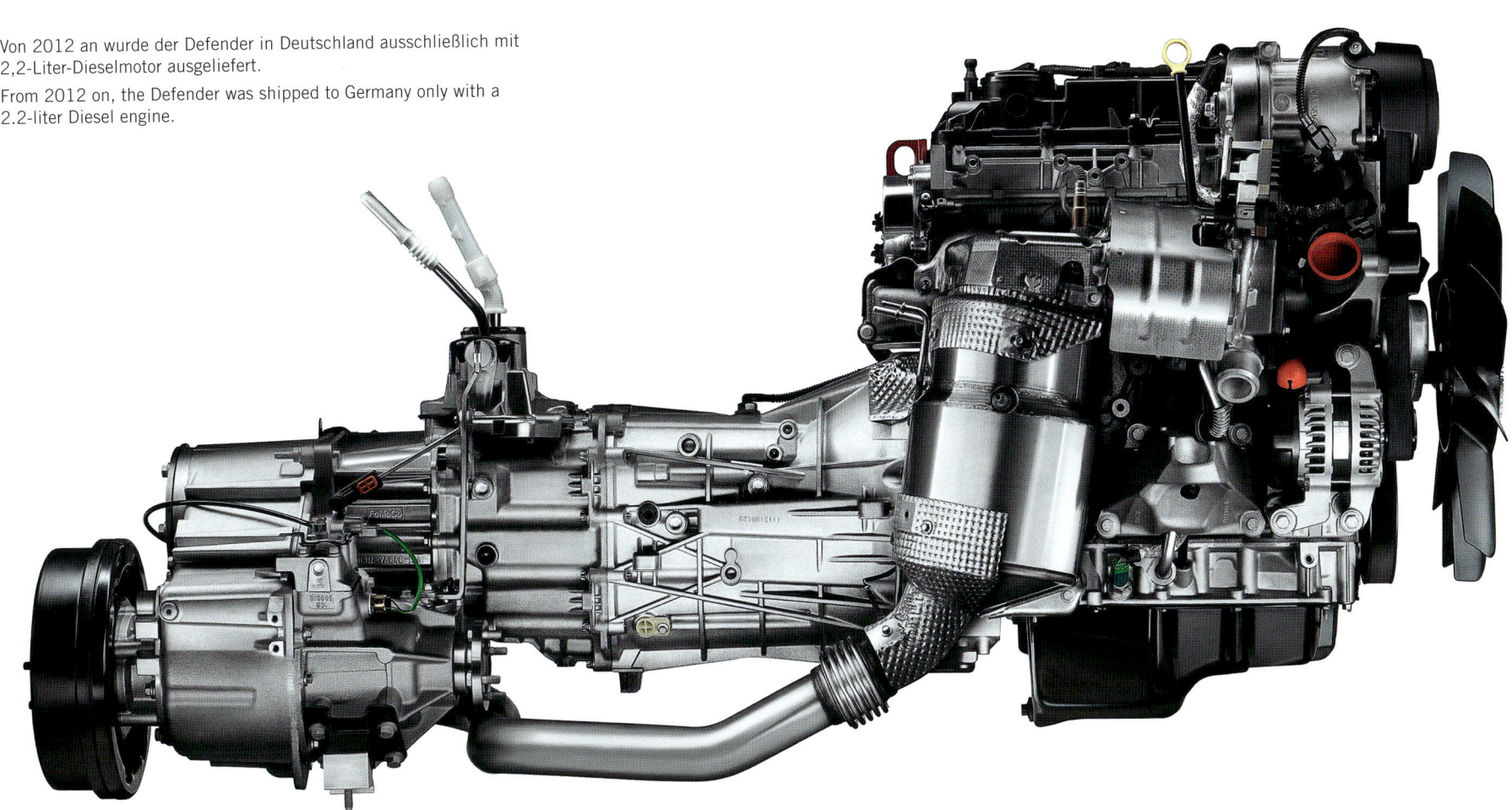

Der Land Rover Defender mutete wie ein Relikt aus einer anderen Zeit an.
Und doch konnte er sich über Jahrzehnte und viele Generationen hinweg
immer wieder neue Fangemeinden erschließen.

The Land Rover looks like a relic from another time. For decades and generations, it was nonetheless always able to find new fans.

Als Mittelklassemodell bietet der XE dennoch typisches Jaguar-Feeling gepaart mit Fahrdynamik.
Despite being a middle class model, the XE offers typical Jaguar feeling and driving dynamics.

Jaguar XE

Fahrzeugdaten

Hersteller:	Jaguar
Land:	Großbritannien
Modell:	XE
Bauzeit:	seit 2015
Länge:	4.672 mm
Breite:	2.075 mm
Höhe:	1.416 mm
Radstand:	2.835 mm
Leergewicht:	1.435 kg
Antriebsart:	Hinterrad
Höchstgeschwindigkeit:	227 km/h
Verbrauch:	3,8 Liter/100 km (Diesel)

*Werte für Jaguar XE E-Performance

Jaguar, vor allem für seine Wagen der Ober- und Spitzenklasse bekannt, startete mit dem im Herbst 2014 vorgestellten XE sein neues Engagement in der Mittelklasse. Dem Jaguar-typischen Aussehen blieb man dabei treu. Das typische Jaguar-Feeling wurde nun auch für kleineres Geld erlebbar. Der XE basiert auf einer neuen modularen Jaguar-Land-Rover-Aluminium-Plattform. Dreiviertel der Karosserie werden in Aluminium-Leichtmetallbauweise realisiert, darunter die Doppelquerlenker und die Mehrlenkerachse. Der hohe Leichtmetallanteil half nicht nur, den Wagen leichter zu machen, sondern auch für eine ausgewogene Gewichtsverteilung zwischen Vorder- und Hinterachse zu sorgen. Im XE feierte zudem eine elektromechanische Servolenkung Premiere. Der Jaguar XE steht in Konkurrenz zu deutschen Automarken und ist diesen in punkto Fahrverhalten durchaus ebenbürtig. Angeboten werden Vierzylinder-Dieselmotoren, 2-Liter-Vierzylinder-Benziner sowie ein 3-Liter-V6.

Specifications

Manufacturer:	Jaguar
Country:	Great Britain
Model:	XE
Produced:	since 2015
Length:	4672 mm
Width:	2075 mm
Height:	1416 mm
Wheelbase:	2835 mm
Empty weight:	1435 kg
Type of drive:	Rear-wheel drive
Max. speed:	227 km/h
Fuel consumption:	3.8 l/100 km (Diesel)

* data for the Jaguar XE E-Performance

Jaguar is mostly known for its luxury class and top class cars, but with the XE introduced in fall 2014, the company started its new middle class commitment. The typical Jaguar design has been retained so that Jaguar feeling now can be experienced for less money. The XE bases on the new modular Jaguar/Land Rover aluminum platform. Three quarters of the auto body are built as an aluminum light metal construction, including the twin suspension arm and the multi-link axle. The high amount of light metal doesn't only help to make the car weigh less but also to balance the weight between front axle and rear axle. Furthermore, an electromechanical power steering debutes in the XE. The Jaguar XE competes against German brands and is on par with them when it comes to the driving behavior. The car is available with a 4-cylinder Diesel engine, a 2-liter 4-cylinder gasoline engine and a 3-liter V6 engine.

Der Motor

Motordaten

Bauart:	Viertakt-Diesel-Reihenmotor mit Turbolader
Zylinderzahl:	4
Hubraum:	1.999 cm³
Bohrung:	83 mm
Hub:	92,4 mm
Leistung:	120 kW (163 PS) bei 4.000/min

Im Jaguar XE E-Performance versieht ein Common-Rail-Dieselmotor mit Turbolader und Direkteinspritzung seinen Dienst. Die maximale Leistung des Reihenvierzylinders beträgt 120 kW (163 PS), die er bei 4.000 Touren abgibt. Im Bereich von 1.750 bis 2.500 Touren liegt das höchste Drehmoment von 380 Nm an. Zu den weiteren Merkmalen des Jaguar-Zwei-Liter-Dieselmotors zählt die variable Steuerung der Auslassventile. Damit werden die Abgasemissionen und der Verbrauch positiv beeinflusst. Weiterhin kommt die selektive katalytische Reduktion zum Einsatz, mit der die Stickoxid-Emissionen auf ein Minimum gesenkt und die weltweit strengsten Abgasnormen erfüllt werden. In der Standardausführung überträgt der Jaguar XE die Kraft des Motors über ein Sechsgang-Schaltgetriebe auf die Hinterräder. Aus dem Stillstand beschleunigt der Wagen binnen 8,4 Sekunden auf 100 km/h. Der Jaguar XE ist auch mit einem in der Hubraumzahl identischen, aber mit 132 kW (180 PS) stärkeren Dieselmotor erhältlich.

The engine

Engine specifications

Type:	Four-stroke Diesel in-line engine with turbo charger
Number of cylinders	4
Displacement:	1999 cm³
Bore:	83 mm
Stroke:	92.4 mm
Power:	120 kW (163 hp) at 4000 rpm

In the Jaguar XE E-Performance, a common rail Diesel engine with turbo charger and direct injection does its duty. The maximum power of the 4-cylinder in-line engine amounts to 120 kW (163 hp) and is reached at 4000 rpm. In the range from 1750 to 2500 rpm, the maximum torque of 380 Nm is present. A further feature of the 2-liter Diesel engine of the Jaguar is the variable outlet valve control, which has a positive effect on emissions and fuel consumption. Selective catalytic reduction reduces the NOx emissions to a minimum and helps to adhere to the world's most restrictive emission standards. In the standard version, the engine power is transferred to the rear wheels via a six-gear transmission. The car can accelerate from 0 to 100 km/h in 8.4 seconds. The Jaguar XE is also available with a Diesel engine, which has the same displacement but is more powerful, delivering 132 kW (180 hp).

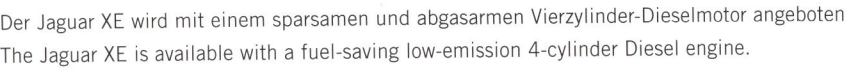
Der Jaguar XE wird mit einem sparsamen und abgasarmen Vierzylinder-Dieselmotor angeboten.
The Jaguar XE is available with a fuel-saving low-emission 4-cylinder Diesel engine.

Mit dem Jaguar XE erfolgte 2015 ein neuer Vorstoß des britischen Automobil-herstellers in der Mittelklasse. Sparsame und leistungsstarke Dieselmotoren passen dabei bestens ins Angebot.

With the Jaguar XE, the British car manufacturer ventures again into the middle class. Fuel-saving and powerful Diesel engines are a perfect supplement to the portfolio.

Zur muskulösen, gedrungenen Formensprache des Camaro passen die exzellenten Leistungsdaten.

The excellent performance data match the muscular and compact shape of the Camaro.

Mit dem „Hyper Concept" wurden 2015 zahlreiche neue Zubehörteile für den Camaro vorgestellt.

With the "Hyper Concept" of 2015, numerous new accessories for the Camaro were advertised.

Chevrolet Camaro

Fahrzeugdaten

Hersteller:	Chevrolet
Land:	USA
Modell:	Camaro
Bauzeit:	seit 2015
Länge:	4.784 mm
Breite:	1.880 mm
Höhe:	1.340 mm
Radstand:	2.812 mm
Leergewicht:	1.539–1.769 kg
Antriebsart:	Hinterrad
Höchstgeschwindigkeit:	240–290 km/h
Verbrauch:	11,1–12,8 Liter/100 km

Im Jahr 2009 ließ General Motors den Camaro in seiner nunmehr fünften Generation wieder auferstehen, und 2015 folgte die Neuvorstellung der mittlerweile sechsten Auflage des legendären US-amerikanischen Klassikers. Grundlage für den als Coupé und von 2016 an auch als Cabriolet erhältlichen Camaro bildete die sogenannte Alpha-Plattform, wie sie bereits beim Cadillac ATS und CTS zum Einsatz kam. Bereits Ende 2015 startete der Verkauf des neuen Performance Cars von Chevrolet. Als neue Basismotorisierung fungiert der Vierzylinder-Reihenmotor Turbo LTG mit 202 kW (275 PS) Leistung, wie er auch im Opel Astra J OPC zum Einsatz kommt. Angesichts von 240 km/h Höchstgeschwindigkeit ermöglicht bereits diese Motorisierung standesgemäße Fahrleistungen. Zudem ist ein 3,6-Liter-V6 mit 246 kW (335 PS) und einem Drehmoment von 385 Nm erhältlich. Zwei besonders kräftige V8-Motoren mit jeweils 6.162 cm³ Hubraum runden die Motorenpalette schließlich nach oben ab.

Specifications

Manufacturer:	Chevrolet
Country:	USA
Model:	Camaro
Produced:	since 2015
Length:	4784 mm
Width:	1880 mm
Height:	1340 mm
Wheelbase:	2812 mm
Empty weight:	1539–1769 kg
Type of drive:	Rear-wheel drive
Max. speed:	240–290 km/h
Fuel consumption:	11.1–12.8 l/100 km

In 2009, General Motors revived the Camaro with a fifth generation. The sixth edition of this legendary US classic followed in 2015. The Camaro is available as a coupé and since 2016 also as a convertible. It is based on the so-called Alpha platform, which was already used in the Cadillac ATS and CTS. Sales of the new performance car by Chevrolet began at the end of 2015. The new base engine is the 4-cylinder in-line engine Turbo LTG with 202 kW (275 hp), which is also used in the Opel Astra J OPC. With a top speed of 240 km/h, even this engine provides an appropriate driving performance. Additionally, a 3.6 liter V6 engine with 246 kW (335 hp) and a torque of 385 Nm is available. The upper end of the engine range is made up of two particularly powerful V8 engines with a displacement of 6162 cm³.

Der Motor

Motordaten

Bauart:	Viertakt-60-Grad-V-Motor
Zylinderzahl:	6
Hubraum:	3.640 cm³
Bohrung:	95 mm
Hub:	85,6 mm
Leistung:	246 kW (335 PS)
	bei 6.800/min

Bereits der 2-Liter-Vierzylinder-Reihenmotor mit 205 kW (275 PS), Direkteinspritzung und variabler Ventilsteuerung zeigt sich dank Turboaufladung als kräftiges Exemplar, er entfaltet bei 3.000 Umdrehungen pro Minute ein Drehmoment von satten 400 Nm. Zudem stellt er 90 Prozent seines maximalen Drehmoments bereits ab 2.100 Touren bereit. Mit dem zum Zeitpunkt der Drucklegung dieses Buches hierzulande offiziell nicht angebotenen 3,6-Liter-V6-Motor steht für den Camaro ein noch potenteres Kraftpaket zur Verfügung. Es leistet 246 kW (335 PS) bei 6.800 Umdrehungen pro Minute und bringt es auf ein maximales Drehmoment von 385 Nm bei 5.300/min. Die beiden Zylinderreihen des V6-Motors sind im Bankwinkel von 60 Grad zueinander angeordnet. Vierventiltechnik sowie variable Ventilsteuerung und Direkteinspritzung weisen dieses Triebwerk als hochmodernen Motor aus, der hohe Leistung mit sehr guter Effizienz bei exzellenten Fahrleistungswerten in Einklang bringt.

The engine

Engine specifications

Type:	4-stroke 60° V-engine
Number of cylinders:	6
Displacement:	3640 cm³
Bore:	95 mm
Stroke:	85.6 mm
Engine power:	246 kW (335 hp)
	at 6800 rpm

Thanks to its turbocharger, even the 2-liter 4-cylinder in-line engine with 205 kW (275 hp), direct injection and variable valve control proves to be a powerful beast: at 3000 rpm, it provides a torque of a whopping 400 Nm. 90 % of its maximum torque are already available at 2100 rpm. Even more power is available with the 3.6 liter V6 engine of the Camaro. It provides 246 kW (335 hp) at 6800 rpm with a maximum torque of 385 Nm at 5300 rpm. The cylinder banks of the V6 engine are placed at an angle of 60°. With its four-valve technology, variable valve control and direct injection system, it is a state-of-the-art engine, which combines high power, high efficiency and excellent drive performance figures.

Der 3,6-Liter-V6-Motor verfügt über Direkteinspritzung sowie eine variable Ventilsteuerung.

The 3.6-liter V6 engine comes with direct injection and variable valve control.

Motoren der sechsten Camaro-Generation: V8, V6 und Reihenvierzylinder mit Turboaufladung (v.li.).

Engines of the sixth Camaro generation: V8, V6 and in-line four-cylinder engine with turbocharger (left to right).

Mit dem neuen Camaro ließ Chevrolet die Legende wieder auferstehen. Ihre sogenannte Alpha-Plattform teilt sie sich in der aktuellen Version mit Fahrzeugen aus dem Hause Cadillac.

Chevrolet revived the legend with the new Camaro. The company shares the current version of the Alpha platform with Cadillac vehicles.

Bildverzeichnis / Photo credits

Audi AG: Umschlag, 30, 33, 54, 57, 70, 73, 78, 81, 90, 93, 94, 97, 98, 101, 102, 105, 106, 109, 194, 197, 202, 205, 238, 241, 294, 297, 322, 325

Daimler AG: 10, 13, 14, 17, 18, 21, 22, 25, 26, 29, 34, 37, 50, 53, 62, 65, 66, 82, 85, 86, 89, 158, 161, 226, 229, 270, 273

Peugeot Kommunikation: 38, 41

FCA US LLC: 42, 45, 46, 49, 142, 145

BMW AG: 58, 61, 130, 133, 154, 157, 218, 221, 234, 237, 246, 249, 274, 277

Fiat Chrysler Automobiles FCA: 74, 77, 182, 185, 258, 261, 278, 281

Citroën Communication/Georges Guyot: 110, 113, 122, 125

Fiat Chrysler Automobiles N.V.: 114, 117

Dr. Ing. h.c. F. Porsche AG: 118, 121, 162, 165, 318, 321

Volkswagen AG: 126, 129, 149, 230, 233, 290, 293

Volvo Car Group: 134, 137, 210, 213, 222, 225

General Motors USA: 138, 141, 190, 193, 306, 309, 338, 341

www.shutterstock.com: 146

Renault: 150, 153, 170, 173

Ford Motor Company: 166, 169